Springer Complexity

Springer Complexity is a publication program, cutting across all traditional disciplines of sciences as well as engineering, economics, medicine, psychology and computer sciences, which is aimed at researchers, students and practitioners working in the field of complex systems. Complex Systems are systems that comprise many interacting parts with the ability to generate a new quality of macroscopic collective behavior through self-organization, e.g., the spontaneous formation of temporal, spatial or functional structures. This recognition, that the collective behavior of the whole system cannot be simply inferred from the understanding of the behavior of the individual components, has led to various new concepts and sophisticated tools of complexity. The main concepts and tools – with sometimes overlapping contents and methodologies – are the theories of self-organization, complex systems, synergetics, dynamical systems, turbulence, catastrophes, instabilities, nonlinearity, stochastic processes, chaos, neural networks, cellular automata, adaptive systems, and genetic algorithms.

The topics treated within Springer Complexity are as diverse as lasers or fluids in physics, machine cutting phenomena of workpieces or electric circuits with feedback in engineering, growth of crystals or pattern formation in chemistry, morphogenesis in biology, brain function in neurology, behavior of stock exchange rates in economics, or the formation of public opinion in sociology. All these seemingly quite different kinds of structure formation have a number of important features and underlying structures in common. These deep structural similarities can be exploited to transfer analytical methods and understanding from one field to another. The Springer Complexity program therefore seeks to foster cross-fertilization between the disciplines and a dialogue between theoreticians and experimentalists for a deeper understanding of the general structure and behavior of complex systems.

The program consists of individual books, books series such as "Springer Series in Synergetics", "Institute of Nonlinear Science", "Physics of Neural Networks", and "Understanding Complex Systems", as well as various journals.

Understanding Complex Systems

Series Editor

J.A. Scott Kelso
Florida Atlantic University
Center for Complex Systems
Glades Road 777
Boca Raton, FL 33431-0991, USA

Understanding Complex Systems

Future scientific and technological developments in many fields will necessarily depend upon coming to grips with complex systems. Such systems are complex in both their composition (typically many different kinds of components interacting with each other and their environments on multiple levels) and in the rich diversity of behavior of which they are capable. The Springer Series in Understanding Complex Systems series (UCS) promotes new strategies and paradigms for understanding and realizing applications of complex systems research in a wide variety of fields and endeavors. UCS is explicitly transdisciplinary. It has three main goals: First, to elaborate the concepts, methods and tools of self-organizing dynamical systems at all levels of description and in all scientific fields, especially newly emerging areas within the Life, Social, Behavioral, Economic, Neuro- and Cognitive Sciences (and derivatives thereof); second, to encourage novel applications of these ideas in various fields of Engineering and Computation such as robotics, nanotechnology and informatics, third, to provide a single forum within which commonalities and differences in the workings of complex systems may be discerned, hence leading to deeper insight and understanding. UCS will publish monographs and selected edited contributions from specialized conferences and workshops aimed at communicating new findings to a large multidisciplinary audience.

A. Kleidon R.D. Lorenz (Eds.)

Non-equilibrium Thermodynamics and the Production of Entropy

Life, Earth, and Beyond

With a Foreword by Hartmut Grassl

Dr. Axel Kleidon
University of Maryland
Department of Geography and Earth System
Science Interdisciplinary Center
2181 Lefrak Hall
College Park, MD 20742-8225
USA

Dr. Ralph D. Lorenz
Lunar and Planetary Laboratory
University of Arizona
1629 E. University Blvd.
Tucson, AZ 85721-0092 USA

ISBN 3-540-22495-5 Springer Berlin Heidelberg New York

Library of Congress Control Number: 2004108637

Springer is a part of Springer Science+Business Media
springeronline.com

Printed in Germany

Typesetting and final processig by PTP-Berlin Protago-TeX-Production GmbH, Germany
Cover design: Erich Kirchner, Heidelberg

Printed on acid-free paper 54/3141/Yu - 5 4 3 2 1 0

Foreword

For many millions of years the Earth has been a life-supporting planet with on average increasing biodiversity and its mean near surface air temperature varying only by a few percent (± 5 Kelvin) around the present mean of about 288K. However, despite this comparably small temperature change, the concentration of a major radiatively active gas, carbon dioxide, was more than double the present anthropogenically enhanced value before glaciation set in and only slightly above half the present value during maximum glaciation, the continents have changed shape and have moved to different geographical latitudes, and the luminosity of the sun has increased substantially.

Which processes have guaranteed this impressive temperature stability? A first candidate with the buffering capacity needed is planetary shortwave albedo, which – by decreasing only from 30 to 29 percent – could cause a radiative forcing of the same magnitude but with opposite sign as a drop in carbon dioxide concentration from its value in an interglacial, like our Holocene, to a typical maximum glaciation value of slightly less than 200 part per million by volume. As the maximum contribution to planetary albedo stems from tropospheric clouds both in the tropics and mid-latitudes, their change could be the key stabilizing agent. But why should cloud cover and/or cloud optical depth increase in an interglacial as compared to the glacial? At present we do not know. Because clouds are the expression of an important diabatic process – phase fluxes of water – these fluxes contribute strongly to entropy production in the atmosphere, second only to longwave radiative flux divergence, which is again strongly modulated by clouds.

For me this book is an exceptional one, as it offers a way forward, maybe the solution. It gives as strong hope that an integral principle, maximum entropy production (MEP), is at work in all open systems with large distance to thermodynamic equilibrium, i.e. those governed by non-linear thermodynamics like the Earth. The low import of entropy, expressed as net shortwave flux density divided by the sun's blackbody radiation (~6000 K) and the high export of entropy, expressed as the net longwave flux density at the top of the atmosphere divided by a typical terrestrial temperature (250 to 300 K), point to strong entropy production within the Earth system. It is largely due to the well-known diabatic processes radiative flux divergence, phase changes of water, turbulent sensible heat flux, and dissipation of turbulent kinetic energy.

This book contains, in addition to these purely physical processes, attempts to integrate life as it enhances diabatic processes through evapotranspiration, higher surface roughness and higher emissivity. Life intensifies the global cycles of water, carbon and nitrogen. If all thermodynamic systems far from equilibrium are subject to MEP, life on Earth included, it would also be a governing principle for the evolution of the Earth system. There would no longer be the need for ad-hoc assumptions, like the Gaia hypothesis. On the contrary, we would have a powerful tool to ask climate and Earth system models – the latter just emerging – what kind of human behaviour would lead to which state as we would be able to add MEP as a constraint in addition to the well-known physical laws and boundary conditions (dynamical, thermodynamical and radiation principles; spectral solar irradiance). These models would then search for the most probable future state which will be attained with very high probability. We could for example also see the consequences of land use changes, including the redistribution of water, which strongly impact biodiversity and the carbon cycle, as well as those changes caused by an enhanced greenhouse effect. This would help us to find a sustainable development path. Additionally, the regional, and perhaps global, consequences of air pollution would become visible.

The MEP principle is also connected to self-organized criticality. It could thus become a tool to better understand the abrupt changes of thermohaline circulation and also local-scale phenomena like avalanches. Besides answers to questions raised earlier, it may even offer means to determine bounds for the best place of a planet with respect to its sun and the composition of its crust best suited for the development of life.

If discrepancies emerge between observations and such diverse modeling for recent history, this tells us about either the lack of information to describe the system or insufficient, maybe incorrect constraints or deficiencies in the handling of diabatic processes.

Earth system or climate models applying or exploring the MEP principle will be extremely demanding of computer time. Thus simplified models will be useful tools in the near future as also demonstrated in this book. Their results, although promising, are still not the real test that MEP governs climate and the Earth system. However, a joint activity of high performance computing centres working with Earth system and climate models could rapidly bring us closer to reality if the global observing system is adequate for a real check. I propose an international basic research project devoted to MEP and Climate, initiated by the group that has been gathered to write the chapters of this book, and which could form the nucleus for basic research with immediate repercussions for the global society. I recommend besides individual research projects a joint action by the World Climate Research Programme (WCRP) and the International Geosphere-Biosphere Programme (IGBP) through the Working Group on Coupled Modelling (WGCM) and the Global Analysis, Integration and Modelling (GAIM) element, respectively; because this kind of research needs global data sets from several disciplines, access to largest computers and best models. At the same time the MEP principle will facilitate the search for better parameterizations as all processes in open systems would also obey it.

It was a great pleasure for me to read all the chapters. I hope that scientists from many different disciplines pick up the chapters most relevant for their future work.

Max-Planck-Institut für Meteorologie
Hamburg, Germany

Hartmut Grassl
Director

Preface

"A theory is more impressive the greater the simplicity of its premises, the more different are the kinds of things it relates, and the more extended its range of applicability. Therefore, the deep impression which classical thermodynamics made on me. It is the only physical theory of universal content, which I am convinced, that within the framework of applicability of its basic concepts will never be overthrown."

Albert Einstein (1879-1955)[1]

This book arose from an encounter between the two editors, a Geography professor and a planetary scientist, two people who might otherwise have little in common. Both of us had independently, along with many of the contributors to this volume, grown aware of the profound importance of nonequilibrium thermodynamics and the potential utility of the principle of Maximum Entropy Production. The possible applications span a bewildering diversity of fields, and thus we felt it useful to all of us to draw some of these threads together in a reference volume that captures the 'state of the art'.

But our encounter at the American Geophysical Union meeting in San Francisco in December 2002 would not have led to our undertaking this book were it not for a growing informal network of researchers in MEP – many of us each feeling alone in the wilderness of our own fields. This network has grown, and many of the ideas in the chapters of this book have been developed at informal workshops, notably a workshop on Maximum Entropy Production at INRA in Bordeaux in April 2003 organized by Roderick Dewar and a series of 'Beyond Daisyworld' workshops organized by Tim Lenton and Inman Harvey. These workshops take considerable time and effort to organize, and the editors therefore are most grateful to these 'unsung heroes' of the field, who as well as bringing MEP researchers together play a vital role in exposing others to the idea.

We thank Christian Caron at Springer Verlag for his encouragement and assistance with this project. We are also most grateful to the contributors to this volume, for their patient hard work in dealing with the editing pro-

[1] quoted in MJ Klein (1967) Thermodynamics in Einstein's Universe. Science 157: 509-516.

cess and the frustrations of document templates. Last, but not least, we are grateful to Ma-Li Kleidon for her help with editing the book chapters.

We hope that with this book we demonstrate the wide potential applicability of thermodynamic concepts, and the principle of Maximum Entropy Production in particular, ranging from the evolution of the Universe, planetary climate systems, life on Earth, and the economic activity of humans and its interaction with the environment.

College Park, Tucson
April 2004

Axel Kleidon
Ralph Lorenz

Contents

List of Contributors

Victor R. Baker
Department of Hydrology and Water Resources
University of Arizona
Tucson, AZ, USA
baker@hwr.arizona.edu

David C. Catling
Astrobiology Program
Department of Atmospheric Sciences
University of Washington
Box 351640
Seattle, WA 98195, USA
davidc@atmos.washington.edu

Eric J. Chaisson
Wright Center &
Physics Department
Tufts University
Medford, MA 02155, USA
eric.chaisson@tufts.edu

Peter M. Cox
Hadley Centre for Climate
Prediction and Research
Fitzroy Road
Exeter Devon, EX1 3PB, UK
peter.cox@metoffice.com

Roderick C. Dewar
Unité d'Ecologie Fonctionelle et
Physique de l'Environnement
INRA, BP 81,
33883 Villenave d'Ornon Cedex,
France;
dewar@bordeaux.inra.fr

Klaus Fraedrich
Meteorologisches Institut
Universität Hamburg
Bundesstraße 55
20146 Hamburg, Germany
fraedrich@dkrz.de

Jonathan Gregory
Hadley Centre for Climate
Prediction and Research
Fitzroy Road
Exeter, Devon, EX1 3PB, UK
jonathan.gregory@metoffice.com

Takamitsu Ito
Program in Atmospheres, Oceans
and Climate
Massachusetts
Institute of Technology
Cambridge, MA 02139, USA
ito@ocean.mit.edu

Davor Juretić
Faculty of Natural Sciences,
Mathematics and Education
University of Split,
Split, Croatia;
juretic@pmfst.hr

Axel Kleidon
Department of Geography and Earth System Science
Interdisciplinary Center
2181 Lefrak Hall
University of Maryland
College Park
MD 20742, USA
akleidon@umd.edu

Timothy M. Lenton
School of Environmental Sciences
University of East Anglia
Norwich NR4 7TJ, UK
t.lenton@uea.ac.uk

Charles H. Lineweaver
Department of Astrophysics and Optics
University of New South Wales
Sydney, 2052, Australia
charley@bat.phys.unsw.edu.au

Ralph D. Lorenz
Lunar and Planetary Lab
University of Arizona
Tucson, AZ 85721, USA
rlorenz@lpl.arizona.edu

Hideaki Miyamoto
Department of Geosystem Engineering
University of Tokyo
Tokyo, Japan
hirdy@lpl.arizona.edu

Hisashi Ozawa
Institute for Global Change Research
Frontier Research System for Global Change
Yokohama 236–0001, Japan
ozawa@jamstec.go.jp

Garth W. Paltridge
IASOS, University of Tasmania
GPO Box 252–77, Hobart
Tasmania, Australia
g.paltridge@utas.edu.au

Olivier M. Pauluis
Courant Institute of Mathematical Sciences
New York University
Warren Weaver Hall
251 Mercer St.
New York
NY 10012-1185, USA
pauluis@cims.nyu.edu

Matthias Ruth
Environmental Policy Program
School of Public Policy
University of Maryland
College Park
MD 20742, USA
mruth1@umd.edu

David W. Schwartzman
Department of Biology
Howard University
Washington, DC 20059, USA
dws@scs.howard.edu

Shinya Shimokawa
National Research Institute for Earth Science and Disaster Prevention
Tsukuba 305–0006, Japan
simokawa@bosai.go.jp

Joël Sommeria
Laboratoire des Ecoulements Géophysiques et Industriels
LEGI/Coriolis 21 Avenue des Martyrs
38 000 Grenoble, France;
sommeria@coriolis-legi.org

Thomas Toniazzo
Hadley Centre for Climate Prediction and Research
Fitzroy Road
Exeter, Devon, EX1 3PB, UK
thomas.toniazzo@metoffice.com

Robert E. Ulanowicz
University of Maryland Center for Environmental Science
Chesapeake Biological Laboratory
Solomons, MD 20688–0038 USA
ulan@cbl.umces.edu

Michael J. Zickel
University of Maryland Center for Environmental Science
Chesapeake Biological Laboratory
Solomons, MD 20688–0038 USA
zickel@earthlink.net

Paško Županović
Faculty of Natural Sciences, Mathematics and Education
University of Split
Split, Croatia
pasko@pmfst.hr

1 Entropy Production by Earth System Processes

Axel Kleidon[1] and Ralph Lorenz[2]

[1] Department of Geography and Earth System Science Interdisciplinary Center, 2181 Lefrak Hall, University of Maryland, College Park, MD 20742, USA
[2] Lunar and Planetary Lab, University of Arizona, Tucson, AZ 85721, USA

Summary. Degradation of energy to lower temperatures and the associated production of entropy is a general direction for Earth system processes, ranging from the planetary energy balance, to the global hydrological cycle and the cycling of carbon by Earth's biosphere. This chapter introduces the application of non-equilibrium thermodynamics to the planetary energy balance of Earth and its neighboring planets. The principles of minimum and maximum entropy production are introduced in the context of Earth system processes. Their applicability to the dynamics of the complex Earth system, such as atmospheric turbulence and the global biotic activity, is outlined. This chapter closes with an overview of the structure of the book and how the chapters relate to the overall theme of non-equilibrium thermodynamics.

1.1 Introduction

Earth system processes perform work by degrading sources of free energy, thereby producing entropy. For instance, the atmospheric circulation is slowed down by friction at the surface, so that it requires continuous input of work to maintain a steady-state circulation. The work to drive the atmospheric circulation is derived from the temperature gradient between the equator and the pole. The associated transport of heat from warmer to colder regions leads to a downgrading of the energy and entropy production. The global hydrological cycle is driven by energy conversions associated with evaporation from a warmer surface into an unsaturated atmosphere and subsequent condensation of water at a cooler temperature in the atmosphere. Finally, life requires sources of free energy to build complex organisms. These three examples do not operate in isolation, but are highly interactive: upward motion in the atmosphere often leads to condensation of water and cloud formation, so that the large-scale patterns of precipitation and uplift are highly correlated. Through its metabolisms, biotic activity substantially affects the chemical composition of the Earth's environment and physical characteristics of the land surface, such as surface albedo and aerodynamic roughness. All these examples involve transformations of energy of different forms, and these transformations are governed by the laws of thermodynamics. This leads us to the question of whether thermodynamics can provide us with meaningful insights into how the steady-state of the complex Earth system operates at a macroscopic, planetary scale.

At the foundation of classical thermodynamics are the first and second laws. The first law formulates that the total energy of a system is conserved, while the second law states that the entropy of an isolated system can only increase. The second law implies that the free energy of an isolated system is successively degraded by diabatic processes over time, leading to entropy production. This eventually results in an equilibrium state of maximum entropy. In its statistical interpretation, the direction towards higher entropy can be interpreted as a transition to more probable states.

However, most systems are not isolated, but exchange energy and/or matter with their environment. For instance, planet Earth exchanges energy with its surroundings by radiation of different wavelengths. The formulations from classical thermodynamics can be applied to non-equilibrium systems which are not isolated (e.g., Prigogine 1962). By exchanging energy of different entropy (or mass) across the system boundary, these systems maintain states that do not represent thermodynamic equilibrium. For these systems, the second law then takes the form of a continuity equation, in which the overall change of entropy of the system dS/dt is determined from the local increase in entropy within the system dS_I/dt and the entropy flux convergence dS_E/dt (i.e., the net flux of entropy across the system boundary):

$$dS/dt = dS_I/dt + dS_E/dt \tag{1.1}$$

In steady state, with no change of the internal entropy S of the system, the production of entropy within the system σ that leads to the increase dS_I/dt balances the net flux of entropy across the system boundary dS_E/dt. The second law in this form then states that $\sigma \geq 0$. A non-equilibrium system can maintain a state of low entropy by "discarding" high entropy fluxes out of the system.

1.2 Entropy Production of Climate Systems

The Earth is a non-equilibrium system in a steady state. At the planetary scale, the absorption of solar radiation is balanced by the emission of terrestrial radiation, leading to the following planetary energy balance:

$$I_0(1 - \alpha_{\mathrm{P}}) - \sigma_{\mathrm{B}} T_R^4 = 0 \tag{1.2}$$

with I_0 being the net flux of solar radiation, α_{P} being the planetary albedo, σ_{B} being the Stefan-Boltzmann constant, and T_R the effective radiative temperature. Using present-day values for Earth of $I_0 = 342\,\mathrm{W\,m^{-2}}$ and $\alpha_{\mathrm{P}} = 0.3$ one obtains a net radiative temperature of $T_R = 255\,\mathrm{K}$.

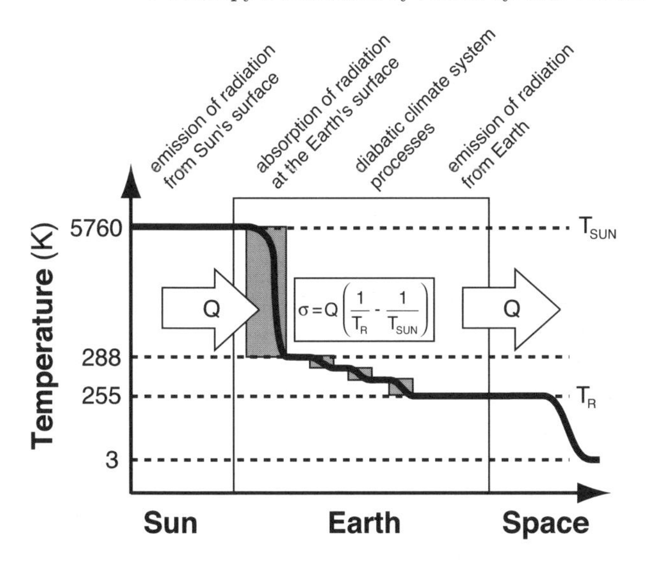

Fig. 1.1. Schematic diagram showing how free energy is subsequently degraded by processes within the Earth system to subsequently lower temperatures

1.2.1 Earth's Climate System

As the flux of energy passes through the Earth system, it is subsequently degraded to lower temperatures, leading to the production of entropy (Fig. 1.1). Solar radiation, emitted at a high radiative temperature of the Sun (of about $T_{SUN} = 5760\,\mathrm{K}$), represents a flux of low entropy. It consists of a flux of photons of high energy, that is, the emitted energy is concentrated on relatively few photons, each carrying a large amount of energy. When solar radiation is absorbed at the Earth's surface at a surface temperature of roughly $T_S \approx 288\,\mathrm{K}$, entropy is being produced in the amount of:

$$\sigma_{\mathrm{RAD}} = Q(1/T_S - 1/T_{SUN}) \tag{1.3}$$

with Q being the amount of radiation being absorbed. Further transformations of the energy take place at subsequently lower temperatures. For instance, the latent heat flux Q_{LH} removes energy from the surface at a temperature T_S and is released to the atmosphere at a lower temperature T_A, leading to entropy production in the amount of:

$$\sigma_{\mathrm{LH}} = Q_{LH}(1/T_A - 1/T_S) \tag{1.4}$$

Similarly, the sensible heat flux, the absorption of terrestrial radiation in the atmosphere, and the transport of heat from warmer to colder regions by the atmosphere and the oceans also contribute to the production of entropy. Ultimately, the absorbed solar radiation is reemitted into space as terrestrial radiation at roughly the radiative temperature T_R. In contrast to solar radiation, this radiation is emitted from Earth at a much lower temperature, representing a flux of photons of less energy and high entropy. The overall entropy production of the Earth system can be estimated from the difference of entropy fluxes across the Earth-space boundary:

$$\sigma_{\mathrm{TOT}} = I_0(1 - \alpha_{\mathrm{P}})(1/T_R - 1/T_{SUN}) \approx 900\,\mathrm{mW\,m^{-2}\,K^{-1}} \tag{1.5}$$

By using the energy fluxes of the global energy balance and estimates for the respective temperatures at which the transformations of energy occur, one can derive the contribution of various diabatic processes of the climate system to the overall production of entropy. Figure 1.2 shows a global estimate of entropy production by different, climate-related processes, based on the analysis of Peixoto et al. (1991) (see also Goody, 2000). From Fig. 1.2 it is evident that the greatest amount of entropy is generated by the absorption of solar radiation in the atmosphere and at the surface. An order of magnitude less is the entropy production associated with the absorption of longwave radiation in the atmosphere and by the transfer of latent heat from the surface to the atmosphere associated with the global hydrological cycle. Seemingly small are the contributions that originate from the sensible heat flux and the frictional dissipation associated with the atmospheric circulation. What Fig. 1.2 shows us is that the overall production of entropy by the Earth system is most profoundly affected by changes in absorption of solar radiation through the planetary albedo, and to a lesser extent by the partitioning of energy at the surface into radiative and turbulent fluxes, and the dynamics of the atmospheric circulation. Since the radiative temperature of the Earth, T_R, is primarily determined by the planetary energy balance, T_R plays a relatively minor role.

1.2.2 Other Planetary Climate Systems

Planets other than the Earth can be considered in an analogous way, as pointed out by Aoki (1983). Catling (this volume) makes the observation that of the planets with atmospheres in our solar system, it is the Earth that has the highest entropy generation rate, perhaps not coincidental with its being an abode for enduring life (Table 1.1). The radiative settings (insolation, albedo and atmospheric opacity) are different for different planetary bodies, yet as discussed later (Sect. 1.3.3 and in Lorenz, this volume) the Maximum Entropy Production principle that appears to drive equator-to-pole heat transport on Earth appears to apply at least to Mars and Titan. Much work remains to quantify the entropy budgets that may pertain on other planets – for example, the heat transports on Mars and Neptune's moon Triton are dominated by the latent heat associated with seasonal sublimation and interhemispheric migration of carbon dioxide and nitrogen respectively.

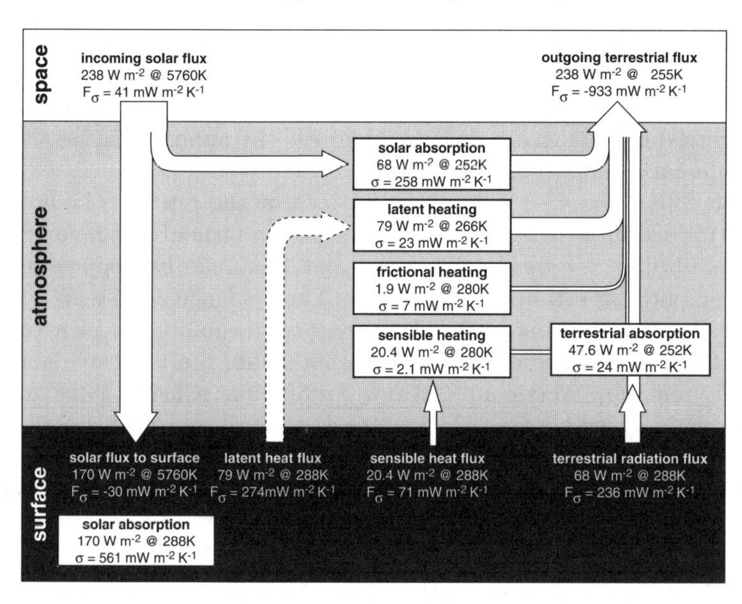

Fig. 1.2. Global budget of entropy production by diabatic Earth system processes. Based on Peixoto et al. (1991). Note that the latent heat flux has been corrected from Peixoto et al.'s estimate

Table 1.1. Planetary entropy production of Earth in comparison to Venus and Mars. Based on data from the National Solar System Data Center (http://nssdc.gsfc.nasa.gov/planetary)

	Venus	Earth	Mars
Luminosity (W m^{-2})	2614	1368	589
Planetary Albedo	0.75	0.31	0.25
Black-body Temperature (K)	232	254	210
Entropy Production (mW m^{-2} K^{-1})·	676	893	507

1.3 The Principles of Minimum and Maximum Entropy Production

Diabatic processes do not produce entropy at an arbitrary rate. Two relevant extremum principles have been formulated to describe the characteristic behavior of non-equilibrium systems. For systems near thermodynamic equilibrium with fixed boundary conditions, Prigogine (1962) formulated the principle of minimum entropy production (MinEP) stating that the steady state of the process is associated with a MinEP state. However, many processes do not have fixed boundary conditions and are far from equilibrium. For those processes it has been shown from information theory (Dewar 2003; also Dewar, this volume) that these processes maintain steady states in which the production of entropy is maximized (MaxEP, or MEP) if there are sufficient degrees of freedom associated with the process (what the specific degrees of freedom are depends on the particular circumstances of the process, see below). Instead of providing formal derivations of the two principles, heat transport is used as an example to highlight the applicability of these two very different principles in the following.

Note that some confusion can arise in how the phrase 'Maximum Entropy Production' is parsed. The attainment of a state of Maximum Entropy (i.e., equilibrium) is a well-accepted one, and Maximum-Entropy methods are widely accepted as estimation techniques. The tendency of systems which are in a steady state, but one that is held away from equilibrium by an external input of energy, to produce entropy at a maximum possible rate, is what we generally mean by 'Maximum Entropy Production'. Clearly there are interesting relationships between the two ideas, as discussed in this book, but it is important to bear the distinction in mind (see also discussion on Shannon information entropy and Maximum Entropy Production in Dewar, this volume).

1.3.1 Heat Transport and Minimum Entropy Production

A simple formulation of heat transport from a warm reservoir with a fixed temperature T_W to a cold reservoir with a fixed temperature T_C can be used to illustrate the MinEP principle. The change of temperature T_M at a location between the two reservoirs is described by the difference of heat fluxes at this location:

$$dT_M/dt = Q_W - Q_C \tag{1.6}$$

with the heat fluxes from the warm reservoir Q_W and to the cold reservoir Q_C expressed as:

$$Q_W = k(T_W - T_M) \tag{1.7}$$

$$Q_C = k(T_M - T_C) \tag{1.8}$$

The rate of entropy production associated with heat transport σ_{HT} is written as:

$$\sigma_{HT} = Q_W(1/T_M - 1/T_W) + Q_C(1/T_C - 1/T_M) \tag{1.9}$$

Figure 1.3 shows the rate of entropy production as a function of T_M. The steady-state is achieved with $Q_W = Q_C$ which leads to zero change of T_M with time (1.6). The steady state is associated with minimum amount of entropy production as shown in Fig. 1.3 (for small $T_W - T_C$). This trivial result can be obtained from the entropy minimization procedure, but since the boundary temperatures are fixed, the result follows from the assumption of steady-state in any case.

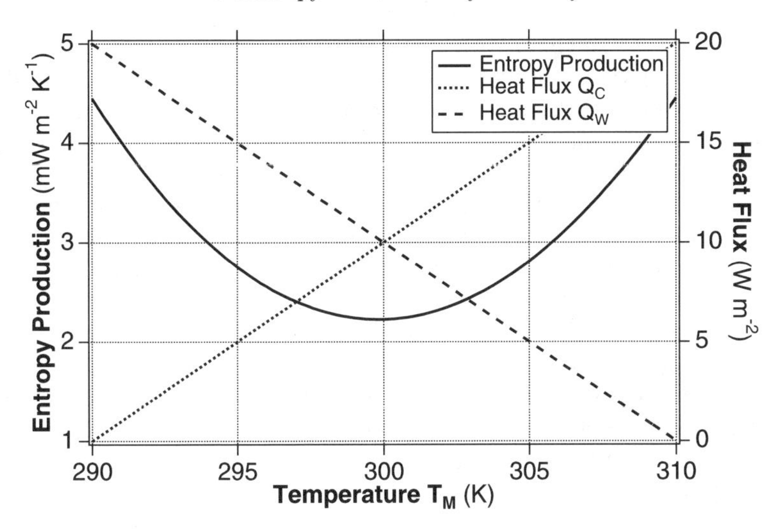

Fig. 1.3. Minimum Entropy Production and heat transport with fixed boundary conditions. Arbitrary conditions of $T_W = 310\,\mathrm{K}$, $T_C = 290\,\mathrm{K}$, and $k = 1$ W m^2 K^{-1} are used

Miyamoto et al. (this volume) describe how a dissipation minimization procedure can be used to form efficient river networks in model hydrological systems (also, Rinaldo et al. 1996). In these systems, the boundary conditions are fixed (equal amount of rain in each cell of a grid: all the flow exits the network through one cell at the corner of the grid so the system is a network which connects all cells to that corner). These fixed boundary conditions make a minimization rather than a maximization the appropriate procedure to apply, and yield networks comparable with (but not quite the same as) river networks observed in nature. We may note that these models are rather analogous to the studies of Bejan (2000), who shows how entropy production minimization can be applied to optimize the design of heat transfer systems (where heat is produced uniformly in a volume or area and must be conveyed to a heat sink at one corner or edge). Although not invoking MinEP, similar scaling relationships to those found for river networks have been derived from fractal geometry for distribution networks of individual, living organisms and plant communities (e.g., Enquist et al. 1998; West et al. 1999).

1.3.2 Heat Transport and Maximum Entropy Production

When we want to describe the heat transport from the tropics to polar regions on Earth, we cannot assume fixed boundary conditions as required by the MinEP principle. The temperatures of the tropics and the poles are determined from the local energy balances, which in turn are affected by the amount of heat that is transported from the tropics to the poles. Also note that k is not a fixed physical property that can be easily measured, since heat transport in the climate system is primarily associated with the turbulent motion of large-scale eddies in the atmosphere and oceans. Therefore, k is determined from the characteristic properties of fluid turbulence in a macroscopic steady state. Under these conditions, it has been proposed that heat transport adjusts itself such that the production of entropy in steady state is at a maximum. The existence of a maximum in entropy production associated with poleward heat transport can easily be demonstrated with a simple energy balance model of the climate system, following earlier work by Lorenz (1960), Paltridge (1975) and others.

We use two boxes to represent the local energy balances of tropical and polar regions and allow for heat transport between the two boxes. The only processes considered are the absorption of solar radiation in the tropics $Q_{IN,T}$ and in the polar regions $Q_{IN,P}$, the emission of terrestrial radiation Q_{OUT}, taken as $a + bT$, and the transport of heat between the boxes Q_{HT}. With these simplifications, we can write the energy balances for the two boxes as

$$Q_{IN,T} - (a + b\,T_T) - Q_{HT} = 0 \tag{1.10}$$

$$Q_{IN,P} - (a + b\,T_P) + Q_{HT} = 0 \tag{1.11}$$

with the empirical coefficients $a = 204\,\mathrm{W\ m^{-2}}$ and $b = 2.17\,\mathrm{W\ m^{-2}\,K^{-1}}$ (Budyko 1969; Sellers 1969).

We can express the heat transport term Q_{HT} as a diffusive flux as:

$$Q_{HT} = k\,(T_T - T_P) \tag{1.12}$$

with k being an effective heat conductivity. The transport of heat from a warmer, tropical reservoir to colder polar regions leads to entropy production σ_{HT}, given by:

$$\sigma_{\mathrm{HT}} = Q_{HT}(1/T_P - 1/T_T) \tag{1.13}$$

In traditional applications of energy balance models (e.g., Budyko 1969; Sellers 1969), the value of k is kept at a fixed value. The MEP hypothesis suggests that the turbulent motion of the atmosphere and oceans adjusts in such a way that σ_{HT} is maximized with respect to k, subject to external constraints such as the conservation of energy, mass and momentum.

Figure 1.4 shows σ_{HT} as a function of k, with a maximum value of σ_{HT} for $k \approx 2\,\mathrm{W\ m^{-2}\ K^{-1}}$. This value is close in magnitude to the commonly used value in energy balance models of North and others (e.g., North et al. 1981). The maximum in entropy production is the result of the competing effects of enhanced heat transport Q_{HT} and reduced temperature gradient $T_T - T_P$ on σ_{HT} with increasing values of k (Fig. 1.4). The planetary rate of entropy production σ_{TOT} also increases with increasing k, since enhanced heat transport leads to a more uniform distribution of temperature, resulting in a reduction of the overall net radiative temperature T_R.

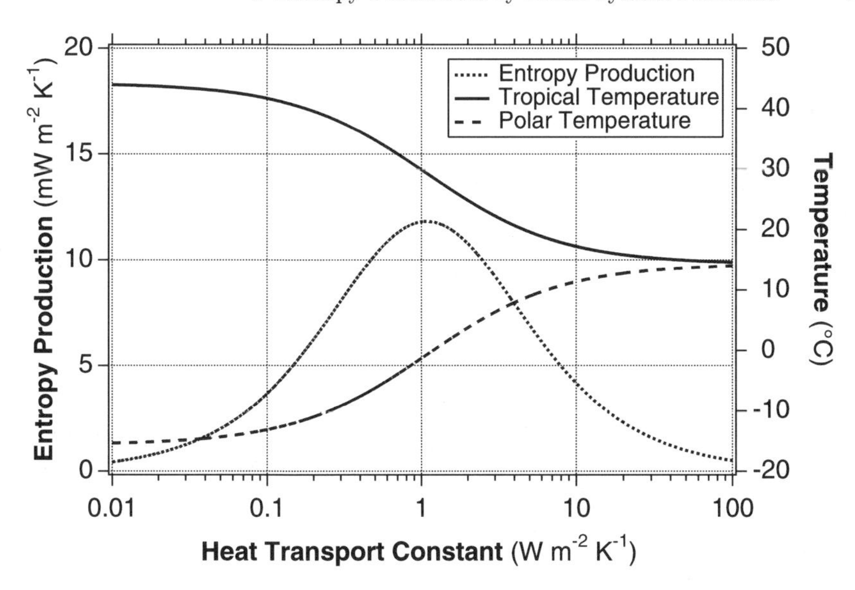

Fig. 1.4. Equator-Pole temperature gradient, and entropy production as a function of heat transport coefficient

At the state of MEP, the atmospheric circulation responds primarily with negative feedbacks to external perturbations. This can be illustrated by considering how the driver of heat transport, that is, the temperature difference between the equator and the pole, reacts to external perturbations. Imagine a fluctuation that leads to a temporary increase in the heat transport coefficient. This leads to a increased warming of the pole at the expense of a stronger cooling in the tropics. Therefore, the equator-pole temperature gradient is reduced, leading to a lower thermodynamic efficiency that uses that heat transport to drive the circulation, resulting in a negative feedback to the perturbation. A similar case can be made for a fluctuation that leads to a temporary decrease in heat transport.

What Lorenz (1960), Paltridge (1975, 1978, 2001) and others (see e.g., review by Ozawa et al. 2003) showed with more detailed energy balance model simulations is that several observed features of the climate system, such as the intensity of the atmospheric circulation, the equator-pole temperature difference in surface temperature, and the meridional distribution of cloud cover reflect a state of the climate system which is close to a state of MEP.

Recently, the MEP principle has been confirmed by simulations with general circulation models that explicitly simulate the fluid dynamics that lead to turbulence (Shimokawa and Ozawa 2001, 2002; Kleidon et al. 2003). Shimokawa and Ozawa (2001, 2002, also this volume) showed with an ocean general circulation model that of the stable steady states of the model, the system assumes the one with the highest rate of entropy production after perturbation. Kleidon et al. (2003) showed with atmospheric general circulation model simulations that increasing the model's spatial resolution increases entropy production up to a certain value after which entropy production is not further increased (also Ito and Kleidon, this volume). They interpreted these results along the lines of Dewar's (2003) interpretation of MEP: spatial resolution affects the spatial degrees of freedom of the atmospheric flow, and therefore entropy production should increase up to the point at which sufficient degrees of freedom are represented in the model.

The confirmation of the MEP principle can be of great utility to Earth system science. Shimokawa and Ozawa's (2001, 2002) results demonstrate that one can assign probability to multiple steady states, and that the MEP state is the most likely state of the climate system. In this context, MEP may help us to constrain possible climatic states of the Earth's past, for which broad-scale observations often lack. Kleidon et al.'s (2003) study motivates the use of MEP as a consistency check for numerical models, in terms of defining a minimum spatial resolution that should be used to adequately simulate large-scale turbulence in the mid-latitude atmosphere, or for tuning simple model parameterizations, such as boundary layer turbulence, for which the degrees of freedom are not explicitly simulated in the model. In fact, if MEP states are not represented by model simulations of the climate systems, that is, that the simulated climate system does not work as hard as it could, it is likely to lead to model biases and misrepresentations of the climate sensitivity to global change (Grassl 1981; Kleidon et al. 2003).

1.3.3 Maximum Entropy Production in a Planetary Context

MEP is of great utility where there is little information to characterize the system's state, particularly of other planets. Lorenz et al. (2001) studied the zonal climates of the planet Mars, and of Saturn's moon Titan with a simple two-box model like that above, and found that the equator-to-pole temperature difference observed is consistent with the MEP hypothesis. Observed temperatures require, and MEP predicts a value of k (often referred to in zonal EBMs as 'D') for Mars that is rather similar to Earth's, and a much smaller value for Titan. These results are rather surprising, in that Mars' atmosphere is very thin, and thus would be expected from dynamical arguments to transport less heat, while Titan's atmosphere is thicker than Earth's and so should transport more heat (especially considering that body's small physical size and the fact that it rotates slowly, a dynamical regime that should encourage efficient equator-to-pole circulations.) MEP does not provide, however, an explanation for these results – it only suggests what the net effect of all transport processes should be. In the case of Mars, it seems that the heat transport can be so large in the thin atmosphere because the latent heat of the Martian CO_2 frost cycle can carry the bulk of the heat required by MEP.

As discussed by Lineweaver (this volume) and Chaisson (this volume), the temperature of the Universe as a whole – the heat sink to which the planets reject heat – has evolved through time. Although the universe was initially prohibitively hot to permit the efficient rejection of heat from evolving systems, the temperature fell to a few tens of K – more or less comparable with the present 3K cosmic background – at the same time that matter began to accumulate into planets. The further evolution of the Universe is based on this temperature differential and allows for the evolution towards higher rates of entropy production and the creation of more complex structures. Certain aspects of this evolution should clearly be reproducible at a macroscopic scale in a thermodynamic sense.

In addition to the bulk transport state of the fluid system, the information-theoretic or probabilistic entropy of the detailed configuration of the system can be a useful exploratory and predictive tool. Sommeria (this volume) shows how applying an entropy maximization procedure to the vorticity distribution of a flowfield such as the circulation of Jupiter's atmosphere can yield remarkable results, such as the emergence of large long-lived vortices like the Great Red Spot.

Thus MEP has a predictive utility for poorly-known environments – only the radiative setting (i.e., insolation and the greenhouse effect) need be known. It may be that the principle is useful for planetary interiors and perhaps for dynamic systems such as planetary rings.

1.3.4 Minimization Versus Maximization of Entropy Production

In summary, it is important to emphasize the differences between the MinEP and the MEP principles. The MinEP principle, as discussed above, applies to linear systems with fixed boundary conditions and few degrees of freedom, and concerns situations where such a system is perturbed away from its steady state. The steady state is one of minimum entropy production relative to any adjacent non-steady state. In contrast, the MEP principle applies to non-linear systems with many degrees of freedom which allow the existence in principle of multiple steady states. The 'chosen' steady state has maximum entropy production relative to the other possible steady states. It is important when dealing with MEP that constraints such as the conservation of energy, mass and momentum are adequately considered in any specific application.

The MEP principle is not without controversy. There have been three prominent objections to MEP raised by meteorologists. The first objection is that there was no apparent justification for the system 'wishing' to find the MEP state. This objection appears to be addressed by the work of Dewar (2003), and the extension of maximum entropy ideas more generally, and by the notion of negative feedbacks associated with MEP states (e.g., Ozawa et al. 2003).

A second objection is that MEP cannot always apply, in that some atmospheres will be too thin to transport the required amount of heat without violating some physical constraint such as the speed of sound (e.g., Rodgers 1976). Rodgers suggested that this would apply to Mercury and Mars. In the latter case, the latent heat of the frost cycle appears to make up for the atmosphere's thinness, and so the objection does not hold. For Mercury, the argument does indeed apply. This objection relates to the conditions laid out above, that MEP is subject to the system constraints.

A similar and more widely-expressed argument is that the heat transport mandated by MEP is independent of the planetary rotation rate. As discussed by Ito and Kleidon (this volume), the constraints associated with the Earth's rotation rate are explicitly considered in climate models and should help to quantify the applicability of MEP. On a rapidly-rotating planet, the circulation required to transport the heat would be dynamically unstable, or would require too much mechanical work generation to sustain against frictional dissipation. Thus, a heat transport lower than that sought by MEP would be observed. However, for very slowly-rotating planets (i.e., where the flow is not constrained by rotational dynamics) MEP should apply, and in any case will specify the maximum possible transport given the constraints of the system.

1.4 Entropy Production and Life on Earth

Every living organism depends on a source of free energy that can be utilized by its metabolism and used to do work (grow, move, reproduce). The basis for every metabolic pathway is a source of free energy provided by the environment. In the words of Schrödinger (1944), life maintains order by degrading free energy and producing high entropy waste (also Boltzmann 1886). Free energy may be derived from geologic sources of chemical compounds (chemotrophs), directly from sunlight (phototrophs), or from organic material (heterotrophs).

1.4.1 Environmental Effects of Biotic Activity

The resources converted by metabolisms are ultimately derived from the environment. Therefore, each metabolic activity necessarily modifies its environment. For instance, photosynthesis converts carbon dioxide into carbohydrates and oxygen, thereby changing the atmospheric concentrations of carbon dioxide and oxygen. Since carbon dioxide plays an important role as an atmospheric greenhouse gas, the rate of photosynthesis indirectly affects the radiative transfer within the atmosphere. The production of some sulfur-related compounds is related to metabolisms deriving their free energy from sulfate oxidation or reduction (or as a byproduct, as is the case for dimethyl sulphide). In the atmosphere, sulfur compounds act as cloud condensation nuclei and impact the rate of formation, the location and brightness of clouds (e.g., Charlson et al. 1987). These examples show the tight linkage between different forms of biotic activity (through their respective metabolisms) and

Table 1.2. Relationships among chemical compounds, biotic activity and atmospheric processes. Also shown are estimates of the global exchange fluxes with the atmosphere related to biotic and abiotic processes under natural conditons

Chemical compound	biotic activity	atmospheric process	biotic exchange flux in 10^{12} g/yr	abiotic exchange flux in 10^{12} g/yr
CO_2	photosynthesis, respiration	absorption of terrestrial radiation	$\approx 210,000^*$	$\approx 600^*$
O_2, O_3	photosynthesis, respiration	absorption of solar radiation in the stratosphere	$\approx 32,000^+$	$\approx 0^+$
CH_4	methanogenesis	absorption of terrestrial radiation	$\approx 150^¶$	$\approx 10^¶$
NO, N_2O NH_3, N_2	nitrogen fixation, nitrification, denitrification	N_2O: absorption of terrestrial radiation	$\approx 155^¶$	$< 3^¶$
SO_2, SO_4, H_2S, DMS	anoxic photo-synthesis, sulfate oxidation and reduction	cloud formation (acts as cloud condensation nuclei)	$\approx 20^¶$	$\approx 162^¶$
H_2O	transpiration by terrestrial vegetation	water cycling (linked with greenhouse effect, cloud formation, heating by latent heat release)	$\approx 35\ 10^{6\,§}$	$\approx 38\ 10^{6\,§}$

* Prentice et al. (2001)
\+ Jacobson et al. (2000)
¶ Schlesinger (1997)
§ Kleidon et al. (2000)

atmospheric functioning at the global scale. Table 1.2 summarizes a few selected chemical compounds and how they relate to biotic activity and atmospheric processes. The last two columns of Table 1.2 provide estimates of the global mass exchange fluxes of these compounds for the atmosphere due to the biota and to abiotic processes. In terms of absolute and relative magnitude, the fluxes of carbon dioxide, oxygen and water are those most strongly modified by biotic activities. These biotic effects lead to a unique composition of the Earth's atmosphere (Table 1.3, also Catling, this volume), with high concentrations of reactive oxygen reflecting a state far from chemical equilibrium.

1.4.2 The Gaia Hypothesis

The strong biotic influence on the atmospheric composition leads to a chemical disequilibrium, most notably reflected in the high concentration of reactive atmospheric oxygen in the Earth's atmosphere (Table 1.3). This chemical disequilibrium is directly linked to the biotic process of photosynthesis that releases oxygen by removing carbon dioxide from the atmosphere. Lovelock (1965) and Hitchcock and Lovelock (1967) suggested to use the existence of such non-equilibrium states of planetary atmospheres as a means to remotely infer the presence of life on other planets.

Based on the notion of Earth's atmospheric composition being far from chemical equilibrium, Lovelock (1972a, b) and Lovelock and Margulis (1974) formulated the controversial Gaia hypothesis, stating "atmospheric homeostasis by and for the biosphere". Associated with the Gaia hypothesis is the notion that biotic feedbacks are primarily negative in nature (which is necessary to maintain homeostasis). Major objections to the Gaia hypothesis include the apparent contradiction to established evolutionary theory, which emphasizes the role of individuals, while the Gaia hypothesis seems to imply a teleological, "goal-seeking" tendency at the planetary scale. Apart from this, it has been pointed out that the hypothesis is ill defined for testing (e.g., Kirchner 1989). The Gaia hypothesis has stimulated a wealth of interdisciplinary research (e.g., Charleston et al. (1987); Schneider and Boston (1991); Schneider et al. (2004)), but is still subject to heated debate (e.g., recent discussion in the journal *Climatic Change*, Kleidon 2002, 2004; Lenton 2002; Lenton and Wilkinson 2003; Volk 2002, 2003a, b; Kirchner 2002, 2003; Lovelock 2003). While the outcome of this debate is yet inconclusive, it is nevertheless important to note that much of the disagreement can be attributed to a difference in perspective, with the planetary perspective promoted by the Gaia hypothesis sharing many similarities with a viewpoint of non-equilibrium statistical mechanics and maximum entropy production. Two contributions in this volume explicitly connect the MEP principle to the Gaia hypothesis (Kleidon and Fraedrich, this volume; Toniazzo et al., this volume). Kleidon and Fraedrich suggest the existence of a MEP state with respect to absorption of solar radiation due to a minimum in the Earth's planetary albedo, which may result from the biotic influence on atmospheric composition and which would have a Gaia-like outcome. Toniazzo et al. address the question of the relation between an MEP climate system and "Gaian" (i.e., climatically active) biota and of the requirements for self-regulation to operate.

1.4.3 Optimization and Entropy Production Within the Biosphere

Apart from the extreme case of the Gaia hypothesis, energy-based principles have been suggested for how ecosystems at a smaller scale organize themselves. These include the notion that ecosystems evolve to maximize the energy flux through the system (Lotka 1922a, b; Loreau 1995), or, formulated

Table 1.3. Atmospheric composition of Earth in comparison to Venus and Mars. Based on data from the National Solar System Data Center (http://nssdc.gsfc.nasa.gov/planetary)

	Venus	Earth	Mars
Surface pressure (bars)	92	1.013	0.0064
Oxygen (%)	≈ 0	20.95	0.13
Carbon dioxide (%)	96.5	0.0350	95.3
Nitrogen (%)	3.5	78	2.7
Water (%)	0.002	≈ 1	0.0210

in slightly different terms, that available energy is degraded at the maximum possible rate, yielding "maximum power" (the maximum power principle of Odum and Pinkerton 1955; Odum 1988). A closely associated observation is that of Ulanowicz and Hannon (1987) who note that vegetated surfaces exhibit cooler surface temperatures and a lower surface albedo, which results in higher rates of entropy production. The trend to increasing entropy production with maturation is interpreted by Ulanowicz and Zickel (this volume) as a trend towards increasing self-organization, which can be quantified by using the concept of ascendency. Schneider and Kay (1994) suggest that ecosystems attain states of maximum dissipation, destroying exergy gradients at a maximum possible rate (with exergy being a measure of the total amount of free energies in the system). This maximization of energy flux should not be seen as only be related to producers (i.e., plants), but in combination with consumption of organic carbon compounds by consumers (e.g., Loreau 1995).

It has also been argued that the emergent optimization of energy fluxes in ecosystems can be understood as the outcome of evolution by natural selection. Alfred Lotka argued in this context that assemblages of organisms can be viewed as "armies of energy transformers", and the "[evolutionary] advantage must go to those organisms whose energy-capturing devices are most efficient in directing available energy into channels favorable to the preservation of the species" (Lotka 1922a, b).

These principles, in one way or another, are dealing with the rate of entropy production, as discussed in the previous sections. They essentially address the question whether a large number of individual organisms organize themselves in any particular state, which connects these principles to a perspective of statistical mechanics and thermodynamics of non-equilibrium systems.

It is important to point out that when we extend the impact of ecosystems on Earth system functioning to the planetary level, boundary conditions are not fixed. The uptake of carbon dioxide by photosynthesis affects the atmospheric composition, so that we deal with a non-linear system, which, through the inherent diversity within the biosphere, should exhibit many degrees of freedom. This line of reasoning suggests that MEP should potentially be applicable to the interactions of the biosphere with its atmospheric environment, just as atmospheric heat transport interacts with the equator-pole temperature gradient (Catling, this volume; Kleidon and Fraedich, this volume). Another way to look at the applicability of MEP to the biosphere is from a perspective of reproducibility. As discussed in Dewar (this volume), one can consider the reproducibility of a macroscopic system state as the key idea behind the second law of thermodynamics. As pointed out by Lineweaver (this volume), this translates to the question of which aspects of the Earth's biosphere are reproducible, in terms of the evolution of Earth's surface temperature and life forms (Schwartzman and Lineweaver, this volume) and in terms of biogeochemical evolution of the atmosphere (Catling, this volume).

1.5 Structure of This Book

The purpose of this book is to provide a general introduction to the statistical mechanics and thermodynamics of non-equilibrium systems and how it applies to Earth system processes, life on Earth in the context of the evolution of the Universe. It is a synthesis of reviews of previous work in combination with recent developments.

The chapters of this book are organized into three parts. The first part focuses on general and theoretical issues of describing the dynamics of complex systems with special emphasis on entropy production. Eric Chaisson (Chap. 2) provides a general introduction of non-equlibrium thermodynamics in a broader context. His chapter outlines the general direction of the evolution of the Universe to higher rates of energy use and rise in general complexity, from the formation of matter to galaxies, life on Earth, to the evolution of complex human societies. The next chapter introduces the principle of Maximum Entropy Production from a historic perspective. Its author, Garth Paltridge, has been central to the development of the MEP principle and how it relates to atmospheric heat transport. Chap. 4 by Roderick Dewar reviews the information-theory based formulation of statistical mechanics as promoted by Jaynes. As a central piece of Dewar's chapter, it is shown how the MEP principle can be derived from information theory, how it connects to macroscopic reproducibility, and how the frequently observed phenomenon of self-organized criticality (SOC) of natural and human systems can also be derived from this perspective. Robert Ulanowicz and Michael Zickel discuss the ecological concept of ascendency in Chap. 5 which can be used to quantify the organization of a complex system (such as an ecosystem). This methodology could be used in physical models to quantify the organization of turbulent flow. In Chap. 6, Charles Lineweaver provides a discussion of the universal limitations to the MEP principle, and which aspects of the evolution of the Universe and the Earth's biosphere are macroscopically reproducible (also Schwartzman and Lineweaver, this volume).

The second part of the book focuses on the application of non-equilibrium thermodynamics and the quantification of entropy production in physical systems. Joël Sommeria describes in Chap. 7 a theroretical derivation of MEP from the fluid dynamics of a two-dimensional system. He discusses the emergent behavior of the system and makes connections to the observed red spot of Jupiter as such an emergent feature. Takamitsu Ito and Axel Kleidon (Chap. 8) deal with entropy production by the atmospheric circulation from a theoretical viewpoint and how it is simulated by atmospheric General Circulation Models (GCMs). They also provide a demonstration of how the MEP principle can be confirmed with a GCM. In Chap. 9, Olivier Pauluis quantifies the rates of entropy production associated with atmospheric moisture, and discusses the role of dissipation associated with water vapor in the Earth's entropy production budget. Shinya Shimokawa and Hisashi Ozawa (Chap. 10) discuss entropy production associated with the oceanic circulation. They demonstrate the existence of multiple steady states with an oceanic GCM and then show that perturbations of these states generally lead to higher rates of entropy production. Hideaki Miyamoto, Victor Baker, and Ralph Lorenz give an overview of the application of thermodynamics to the formation of river networks and emerging scaling laws in chapter 11. In the last chapter of the second part (Chap. 12), Ralph Lorenz discusses the extension of thermodynamics and MEP to phenomena in the solar system and other planets.

The third part of the book deals with the application of thermodynamics to the Earth's biosphere – from a molecular level in Chap. 13 to the Gaia hypothesis (Chap. 17) and the economic activity of the anthroposphere (Chap. 18). In Chap. 13, Davor Juretić and Paško Županović propose the application of thermodynamics to the quantification of rates in initial photosynthetic reactions, pointing out the non-linearities associated with these reactions such that MEP should be applicable. Axel Kleidon and Klaus Fraedrich focus on the large-scale effects of terrestrial vegetation on the physical exchanges of energy and water at the land surface in Chap. 14. They discuss how the MEP principle should be applicable to understand these interactions, and how Gaia-like behavior may result from a MEP state. In Chap. 15, David Catling describes the evolution of the Earth's atmosphere and how it has been affected by the biosphere. He also points out the general energetic advantages of high oxygen concentrations to the biosphere. Chapter 16 by David Schwartzman and Charley Lineweaver covers the biotic evolution on Earth from a viewpoint of temperature constraints and how it is interrelated with global biogeochemical cycles. They argue that the major trends reflected in the early evolution of the Earth's biosphere should be viewed as a deterministic, that is reproducible aspect of biotic evolution on Earth. In Chap. 17, Thomas Toniazzo, Timothy Lenton, Peter Cox and Jonathan Gregory discuss the Gaia hypothesis and how it may relate to the MEP principle. They use the Daisyworld model of Watson and Lovelock (1983) – which was originally developed for demonstrating planetary regulation – to discuss the role of time constants associated with growth for global regulation. The third part of the book closes with chapter 18 by Matthias Ruth. This chapter provides an important linkage to demonstrate the use of thermodynamics to describe the emergent behavior of complex economies. As our Earth system is increasingly dominated by the human species, economic and social processes play an increasing role in shaping the physical environment, and should ultimately affect the rates of entropy production of Earth system processes as well.

As pointed out above, there will doubtless be objections to the MEP principle, notably in determining its applicability domain – when and where is it useful. It is hoped that this volume will encourage that discussion. Ultimately, as MEP becomes a more widely known principle, results will speak for themselves, and help us to provide a general framework to describe the emergent behavior of complex dynamic systems.

References

Aoki I (1983) Entropy production on the Earth and other planets of solar system. J Phys Soc Jpn 52: 1075–1078.

Bejan A (2000) Shape and Structure, from Engineering to Nature. Cambridge University Press, Cambridge, England. 324pp.

Boltzmann L (1886) Der zweite Hauptsatz der mechanischen Wärmetheorie. Sitzungsber Kaiserl Akad Wiss, Wien, Austria.

Budyko MI (1969) Effects of solar radiation variations on climate of Earth. Tellus 211: 611–619.

Charlson RJ, Lovelock JE, Andreae MO, Warren SG (1987) Oceanic phytoplankton, atmospheric sulphur, cloud albedo, and climate. Nature 326: 655–661.

Dewar RC (2003) Information theory explanation of the fluctuation theorem, maximum entropy production, and self-organized criticality in non-equilibrium stationary states. J Physics A 36: 631–641.

Enquist BJ, Brown JH, West GB (1998) Allometric scaling of plant energetics and population density. Nature 395: 163–165.

Goody R (2000) Sources and sinks of climate entropy. Q J Roy Meteorol Soc 126: 1953–1970.

Grassl H (1981) The climate at maximum entropy production by meridional atmospheric and oceanic heat fluxes. Q J R Meteorol Soc 107: 153–166.

Hitchcock DR, Lovelock JE (1967) Life detection by atmospheric analysis. Icarus 7: 149–159.

Jacobson MC, Charlson RJ, Rodhe H, Orians GH (2000) Earth system science – from biogeochemical cycles to global change. Academic Press, San Diego.

Kirchner JW (1989) The Gaia hypothesis: can it be tested? Rev Geophys 27: 223–235.

Kirchner JW (2002) The Gaia hypothesis: Fact, theory, and wishful thinking. Clim Change 52: 391–408.

Kirchner JW (2003) The Gaia hypothesis: conjectures and refutations. Clim Change 58: 21–45.

Kleidon A (2002) Testing the effect of life on Earth's functioning: How Gaian is the Earth system? Clim Change 52: 383–389.

Kleidon A (2004) Beyond Gaia: Thermodynamics of life and Earth system functioning. Clim Change.

Kleidon A, Fraedrich K, Heimann M (2000) A green planet versus a desert world: estimating the maximum effect of vegetation on land surface climate. Clim Change 44: 471–493.

Kleidon A, Fraedrich K, Kunz T, Lunkeit F (2003), The atmospheric circulation and states of maximum entropy production. Geophys Res Lett 30: 2223, doi:10.1029/2003GL018363.

Lenton TM (2002) Testing Gaia: The effect of life on Earth's habitability and regulation. Clim Change 52: 409–422.

Lenton TM, Wilkinson DM (2003) Developing the Gaia theory. Clim Change 58: 1–12.

Loreau M (1995) Consumers as maximizers of matter and energy flow in ecosystems. Am Nat 145: 22–42.

Lorenz EN (1960) Generation of available potential energy and the intensity of the general circulation. in: Pfeffer, R.C. (ed), 'Dynamics of Climate', Pergamon Press, Oxford, UK, pp 86–92.

Lorenz RD, Lunine JI, Withers PG, McKay CP (2001) Titan, Mars and Earth: Entropy production by latitudinal heat transport. Geophys Res Lett 28, 415–418.

Lotka AJ (1922a) Contribution to the energetics of evolution. Proc Natl Acad Sci USA 8: 147–151.

Lotka AJ (1922b) Natural selection as a physical principle. Proc Nat Acad Sci USA 8: 151–154.

Lovelock JE (1965) A physical basis for life detection experiments. Nature 207: 568–570.

Lovelock JE (1972a) Gaia as seem through the atmosphere. Atmospheric Environment 6: 579–580.

Lovelock JE (1972b) Gaia: A new look at life on Earth. Oxford University Press, Oxford.

Lovelock JE (2003) Gaia and emergence – a response to Kirchner and Volk. Clim Change 57: 1–3.

Lovelock JE, Margulis L (1974) Atmospheric homeostasis by and for the biosphere: the Gaia hypothesis. Tellus 26: 2–10.

North GR, Cahalan RF, Coakley JA (1981) Energy balance climate models. Rev Geophys Space Phys 19: 91–121.

Odum HT (1988) Self-organization, transformity, and information, Science 242: 1132–1139.

Odum HT, Pinkerton RC (1955) Time's speed regulator: the optimum efficiency for maximum power output in physical and biological systems. Am Sci 43: 331–343.

Ozawa H, Ohmura A, Lorenz RD, Pujol T (2003) The second law of thermodynamics and the global climate system – A review of the maximum entropy production principle. Rev Geophys 41: 1018.

Paltridge GW (1975) Global dynamics and climate – a system of minimum entropy exchange, Q J R Meteorol Soc 101: 475–484.

Paltridge GW (1978) The steady-state format of global climate. Q J Roy Met Soc 104: 927–945.

Paltridge GW (2001) A physical basis for a maximum of thermodynamic dissipation of the climate system. Q J R Meteorol Soc 127: 305–313.

Peixoto JP, Oort AH, de Almeida M, Tome A (1991) Entropy budget of the atmosphere. J Geophys Res 96: 10, 981–10, 988.

Prentice IC, Farquhar GD, Fasham MJR, Goulden ML, Heimann M, Jaramillo VJ, Kheshgi HS, Le Quéré C, Scholes RJ, Wallace DWR (2001) The carbon cycle and atmospheric carbon dioxide. in: Climate change 2001: The scientific basis. Contribution of working group I to the third assessment report of the Intergovernmental Panel on Climate Change. Houghton JT, Ding Y, Griggs DJ, Noguer M, van der Linden PG, Dai X, Maskell K, Johnson CA (eds). Cambridge University Press, Cambridge, United Kingdom and New York, NY, USA.

Prigogine I (1962) Introduction to Non-equilibrium Thermodynamics.Wiley Interscience, New York.

Rinaldo A, Maritan A, Colaiori F, Flammini A, Rigon R, Rodriguez-Iturbe I, Banavar JR (1996) Thermodynamics of fractal networks. Phys Rev Lett 76: 3364–3367.

Rodgers CD (1976) Minimum entropy exchange principle – reply. Q J Roy Meteor Soc 102: 455–457.

Schlesinger WH (1997) Biogeochemistry: an analysis of global change. 2nd edition, Academic Press, San Diego.

Schneider ED, Kay JJ (1994) Life as a manifestation of the second law of thermodynamics. Math Comput Modeling 19: 25–48.

Schneider SH, Boston PJ (1991) Scientists on Gaia. MIT Press, Cambridge, Mass.

Schneider SH, Miller JR, Crist E, Boston PJ (2004) Scientists debate Gaia: The next century. MIT Press, Cambridge, Mass.

Schrödinger E (1944) What is life? The physical aspect of the living cell. Cambridge University Press, Cambridge, UK.

Sellers WD (1969) A global climate model based on the energy balance of the Earth atmosphere system. J Appl Met 8: 392–400.

Shimokawa S, Ozawa H (2001) On the thermodynamics of the oceanic general circulation: Entropy increase rate of an open dissipative system and its surroundings. Tellus 53A: 266–277.

Shimokawa S, Ozawa H (2002) On the thermodynamics of the oceanic general circulation: Irreversible transition to a state with higher rate of entropy production. Q J Roy Meteorol Soc 128: 2115–2128.

Ulanowicz RE, Hannon BM(1987) Life and the production of entropy. Proc R Soc Lond B 232: 181–192.

Volk T (2002) Towards a future for Gaia theory. Clim Change 52: 423–430.

Volk T (2003a) Seeing deeper into Gaia theory – a reply to Lovelock's response. Clim Change 57: 5–7.

Volk T (2003b) Natural selection, Gaia, and inadvertent by-products: A reply to Lenton and Wilkinson's response. Clim Change 58: 13–19.

Watson AJ, Lovelock JE (1983) Biological homeostasis of the global environment: the parable of Daisyworld, Tellus 35B: 284–289.

West GB, Brown JH, Enquist BJ (1999) The fourth dimension of life: fractal geometry and allometric scaling of organisms. Science 284: 167–169.

2 Non-equilibrium Thermodynamics in an Energy-Rich Universe

Eric J. Chaisson

Wright Center & Physics Department, Tufts University, Medford, MA 02155, USA

Summary. Free energy, the ability to do work, is the most universal currency known in the natural sciences. In an expanding, non-equilibrated Universe, it is free energy that drives order from disorder, from big bang to humankind, in good accord with the second law of thermodynamics and leading to the production of entropy. On all scales, from galaxies and stars to planets and life, the rise of complexity over the course of natural history can be uniformly quantified by analyzing the normalized flow of energy through open, non-equilibrium, thermodynamic systems.

2.1 Introduction

Emerging now from modern science is a unified scenario of the cosmos, including ourselves as sentient beings, based on the time-honored concept of change. Change does seem to be universal and ubiquitous, much as the ancient Greek Heraclitus claimed long ago: "Nothing permanent except change ... all flows." Twenty-five centuries later, evidence for change abounds, some of it obvious, other subtle. From galaxies to snowflakes, from stars and planets to life itself, we are weaving an intricate pattern penetrating the fabric of all the natural sciences – a sweepingly inclusive view of the order and structure of every known class of object in our richly endowed Universe.

Cosmic evolution is the study of the sum total of the many varied developmental and generational changes in the assembly and composition of radiation, matter, and life throughout all space and across all time. These are the physical, biological, and cultural changes that have produced, in turn, our Galaxy, our Sun, our Earth, and ourselves. The result is a grand evolutionary synthesis bridging a wide variety of scientific specialties – physics, astronomy, geology, chemistry, biology, and anthropology, among others – a genuine narrative of epic proportions extending from the beginning of time to the present, from big bang to humankind.

Yet questions remain: How valid are the apparent continuities among Nature's historical epochs and how realistic is this quest for unification? Can we reconcile the observed constructiveness of cosmic evolution with the inherent destructiveness of thermodynamics? Is there an underlying principle, a unifying law, or perhaps an ongoing process that does create, order, and maintain all structures in the Universe, enabling us to study everything on uniform, common ground – "on the same page," sort to speak.

Recent research, guided by notions of unity and symmetry and bolstered by vast new databases, suggests affirmative answers to some of these queries: Islands of ordered complexity – namely, open systems such as galaxies, stars, planets, and life forms that produce entropy to maintain order – are more than balanced by great seas of increasing disorder elsewhere in the environments beyond those systems. All can be shown to be in quantitative agreement with the principles of thermodynamics, especially non-equilibrium thermodynamics. Furthermore, flows of energy engendered largely by the expanding cosmos do seem to be as universal a process in the origin of structured systems as anything yet found in Nature. The optimization of such energy flows might well act as the motor of evolution broadly conceived, thereby affecting all of physical, biological, and cultural evolution (Chaisson 2001).

2.2 Time's Arrow

Figure 2.1 shows an archetypal sketch of cosmic evolution – the "arrow of time." Regardless of its shape or orientation, such an arrow represents an intellectual guide to the sequence of events that have changed systems from simplicity to complexity, from inorganic to organic, from chaos in the early Universe to order more recently. That sequence, as determined by a large body of post-Renaissance data, accords well with the idea that a thread of change links the evolution of primal energy into elementary particles, the evolution of those particles into atoms, in turn of those atoms into galaxies and stars, and of stars into heavy elements, further in turn the evolution of those elements into the molecular building blocks of life, of those molecules into life itself, and of intelligent life into the cultured and technological society that we now share. Despite the compartmentalization of today's academic sciences, evolution knows no disciplinary boundaries.

As such, the most familiar kind of evolution – biological evolution, or neo-Darwinism – is just one, albeit important, subset of a much broader evolutionary scheme encompassing more than mere life on Earth. In short, what Darwinian change does for plants and animals, cosmic evolution aspires to do for all things. And if Darwinism created a revolution in understanding by helping to free us from the notion that humans basically differ from other life forms on our planet, then cosmic evolution extends that intellectual revolution by treating matter on Earth and in our bodies no differently from that in stars and galaxies beyond.

Time's arrow implies no anthropocentrism. It merely provides an intellectual roadmap that symbolically traces increasingly complex structures, from spiral galaxies to rocky planets to reproductive beings. Nor does the arrow mean to imply that "lower," primitive life forms biologically changed directly into "higher," advanced organisms, any more than galaxies physically change into stars, or stars into planets. Rather, with time – much time – environmental conditions suitable for spawning primitive life eventually changed into those favoring the emergence of more complex species; likewise, in the earlier Universe, environments ripe for galactic formation eventually gave way to conditions more conducive to stellar and planetary formation; now, at least on Earth, cultural evolution dominates. Change in environments usually precedes change in systems, and the resulting system changes have *generally* been toward greater amounts of order and complexity.

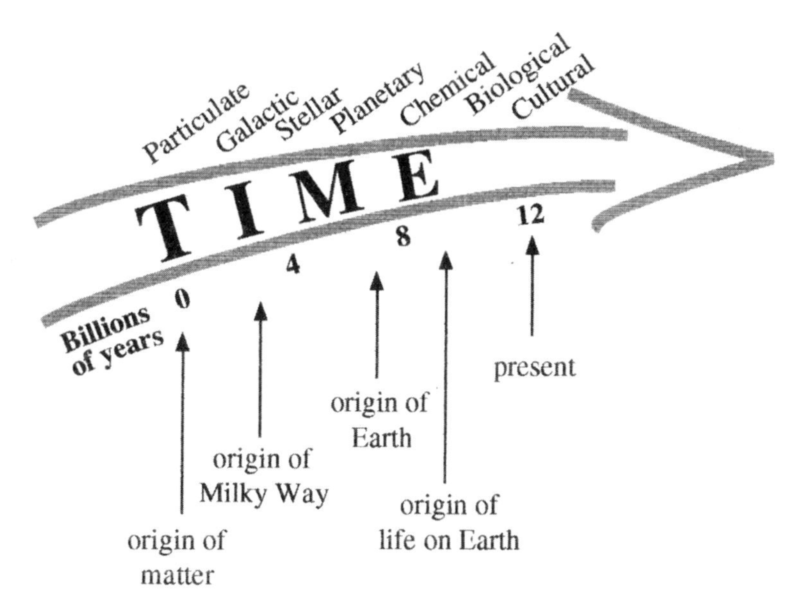

Fig. 2.1. This symbolic "arrow of time" highlights salient features of cosmic history, from its fiery origins some 14 billion years ago (*at left*) to the here and now of the present (*at right*). Labeled diagonally across the *top* are the major evolutionary phases that have produced, in turn, increasing amounts of order and complexity among all material systems: particulate, galactic, stellar, planetary, chemical, biological, and cultural evolution. Cosmic evolution encompasses all these phases. Time is assumed to flow linearly and irreversibly, unfolding at a steady pace, much as other central tenets are assumed, such as the fixed character of physical law or the mathematical notion that $2 + 2 = 4$ everywhere

Figure 2.2 illustrates the widespread impression that material assemblages have become more organized and complex, especially in relatively recent times. This family of curves refers to islands of complexity comprising systems per se – whether giant stars, buzzing bees, or urban centers – not their vastly, increasingly disorganized surroundings. A central task of complexity science aims to explain this temporal rise of organization.

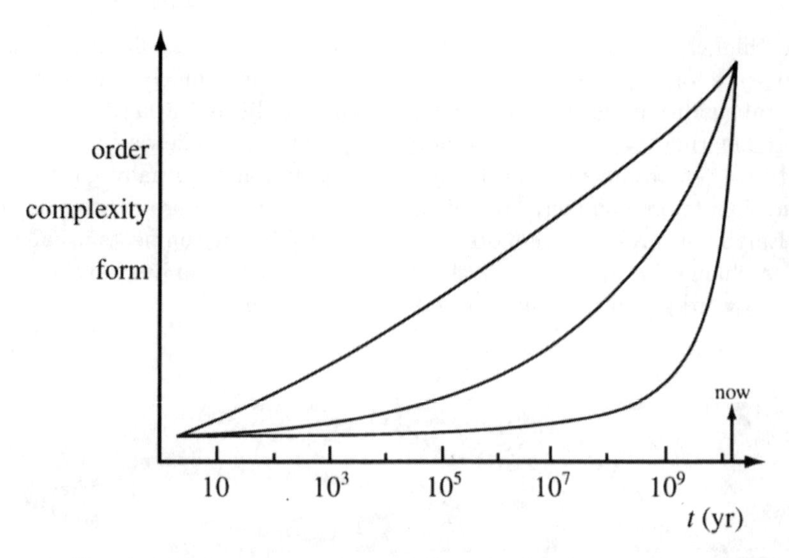

Fig. 2.2. Sketched here qualitatively is the rise of order, form, and structure typifying the evolution of *localized* material systems throughout the history of the Universe. This family of curves connotes the widespread, innate feeling that complexity of ordered structures has *generally* increased over the course of time. Whether this rise of complexity has been linear, exponential, or hyperbolic (as drawn here), current research aims to specify this curve, to characterize it quantitatively. All subsequent graphs in this article have the same temporal scale

2.3 Cosmological Setting

The origin of Nature's many varied structures is closely synonymous with the origin of free energy. Time marches on, equilibrium fails, and free energy flows because of cosmic expansion (Gold 1962; Layzer 1976), all of it summarized by the run of energy densities shown in Fig. 2.3. Here, the essence of change is plotted on the largest scale – the truly big picture, or "standard model," of the whole Universe – so these curves pertain to nothing in particular, just everything in general. They track the main trends, minus devilish details, of modern cosmology: the cooling and thinning of radiation and matter, largely based on observations of distant receding galaxies and of the microwave background radiation – all this change fundamentally driven by the expansion of the Universe.

Radiation completely ruled the early Universe. Life was then non-existent and matter itself only a submicroscopic precipitate suspended in a glowing fireball of intense light, x rays, and gamma rays. Structure of any sort had yet to emerge; the energy density of radiation was too great. If single protons tried to capture single electrons to make hydrogen atoms, radiation was then so fierce as to destroy those atoms immediately. Prevailing conditions during the first few tens of millennia after the origin of time were uniform, symmetrical, equilibrated, and boring. We call it the Radiation Era.

Eventually and inevitably, as also depicted in Fig. 2.3, the primacy of radiation gave way to matter. As the expanding Universe naturally cooled and thinned, charged particles assembled into neutral atoms, among the simplest of all structures; the energy density of matter began to dominate. This represents a change of first magnitude – perhaps the greatest change of all time – for it was as though an earlier, blinding fog had lifted; cosmic uniformity was punctured, its symmetry broken, its equilibrium gone perhaps forever. The Universe thereafter became transparent, as photons no longer scattered aimlessly and destructively. The bright Radiation Era gradually transformed into the darker Matter Era about 10^5 years after the big bang, which is when the free energy began to flow.

Thermodynamics tells us not what will happen, only what *can* happen. This analysis suggests that changing environmental conditions gave rise to the *potential* growth of order and structure. Once symmetry broke and equilibrium failed a few thousand centuries after the start of all things, the temperatures of matter and of radiation diverged with time; thereafter gradients were naturally established owing to cosmic expansion. And this apparently did lead to order among localized systems able to select and utilize, perhaps optimally, the available free energy, resulting in a trend of increasing rates of entropy production (also Lineweaver, this volume).

Figure 2.4 graphs the run of entropy, S, for a thermal gradient typical of a heat engine, here for the whole Universe. This is not a mechanical device running with idealized Newtonian precision, but a cosmological setting potentially able to do work as locally emerging systems interact with their environments – especially those systems able to take advantage of increasing flows of free energy resulting from cosmic expansion and its naturally growing gradients. Although thermal and chemical (but not gravitational) entropy must have been maximized in the early Universe, hence complexity in the form of any structures then non-existent, the start of the Matter Era saw the environmental conditions become more favorable for the potential growth of order, taken here as a "lack of disorder." At issue was timing: As density ρ decreased, the equilibrium reaction rates ($\propto \rho$) fell below the cosmic expansion rate ($\propto \rho^{1/2}$) and non-equilibrium states froze in. Thus we have a seemingly paradoxical yet significant result that, in an expanding Universe, both the disorder (i.e., net entropy) and the order (maximum possible entropy minus actual entropy at any given time) can increase simultaneously – the former globally and the latter locally. All the more interesting when comparing the shape of this curve of potentially increasing order ($S_{max} - S$) in Fig. 2.4 with our earlier intuited sketch of rising complexity in Fig. 2.2.

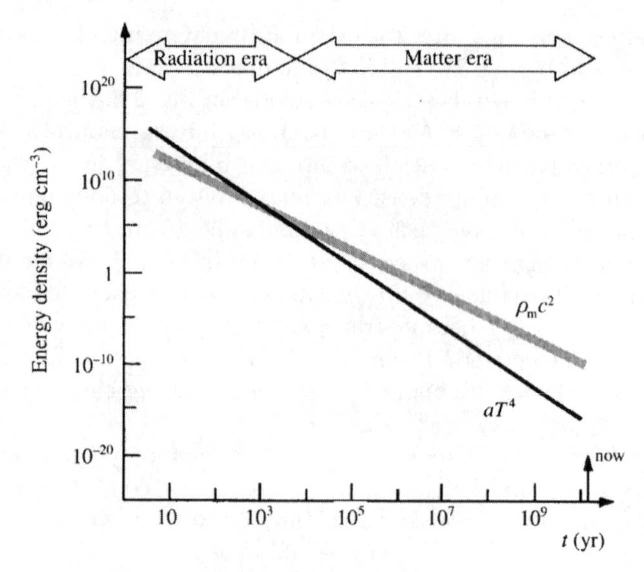

Fig. 2.3. The temporal behavior of both matter energy density (ρc^2) and radiation energy density (aT^4) illustrates perhaps the greatest change in all of history. Here, ρ is the matter density, c the speed of light, a the radiation constant, and T the temperature. Where the two curves intersect, neutral atoms began to form. By some 10^5 years, the Universe had changed greatly as thermal equilibrium and particle symmetry had broken, and the Radiation Era transformed into the Matter Era. A uniform, featureless state characterizing the early Universe thus naturally became one in which order and complexity were thereafter possible. The thicker width of the matter density curve represents the range of uncertainty in total mass density, whose value depends on the (as yet unresolved issue of) "dark matter." By contrast, the cosmic background temperature is well measured today, and its thin curve can be accurately extrapolated back into the early Universe. The startling possibility, recently discovered, that universal expansion might be accelerating should not much affect these curves to date

2.4 Complexity Rising

Complexity, like its allied words *time* and *emergence*, is a term easily spoken yet poorly defined. Although used liberally throughout today's scientific community, complexity eludes our ability to characterize it or to measure it, let alone to specify its true meaning. Complexity: "a state of intricacy, complication, variety, or involvement, as in the interconnected parts of a system – a quality of having many interacting, different components." But what does that mean, scientifically? And can we quantify it, much as for radiation and matter above?

Researchers from many disciplines now grapple with the term *complexity*, yet their views are often restricted to their own specialties, their focus non-unifying; few can agree on either a qualitative or quantitative use of the term.

Some, for example, aspire to model biological complexity in terms of non-junk genome size (Szathmary and Smith 1995); others prefer morphology and flexibility of behavior (Bonner 1988); still others cite numbers of cell types (Kauffman 1993), cellular specialization (McMahon and Bonner 1983), or even physical sizes of organisms per se. Using fluid flow, such as energy, as a basis, Ulanowicz and Zickel (this volume) suggest another method to specify organization and complexity of a system. However, few of these attributes are easily quantified, fewer still serve to measure complexity broadly.

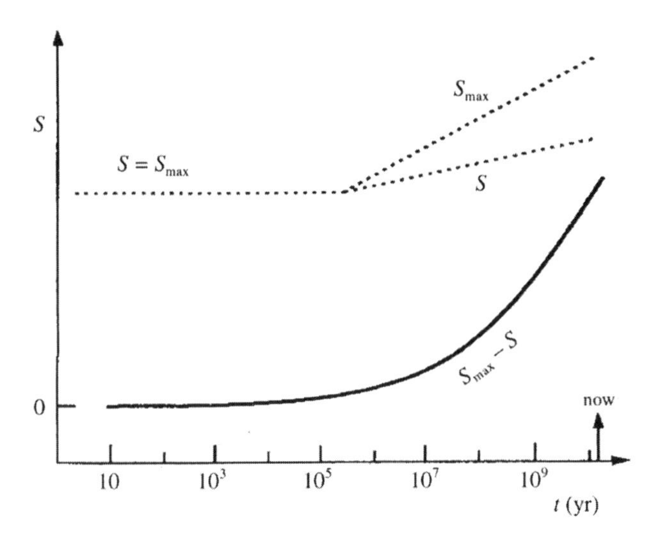

Fig. 2.4. In the expanding Universe, the actual entropy, S, increases less rapidly than the maximum possible entropy, S_{max}, once the symmetry of equilibrium broke when matter and radiation decoupled at $\sim 10^5$ years. By contrast, in the early, equilibrated Universe, $S = S_{max}$ for the prevailing conditions. The potential for the growth of order – ($S_{max} - S$), shown as the thick black curve – has increased ever since the start of the Matter Era. Accordingly, the expansion of the Universe can be judged as the ultimate source of free energy, promoting the evolution of order in the cosmos. This potential rise of order compares well with the family of curves of Fig. 2.2 and provides a theoretical basis for the growth of systems complexity

Putting aside as unhelpful the idea of information content (of the Shannon-Weaver type, which is controversial even if sometimes useful) and of negative entropy (or "negentropy," which Schroedinger (1944) first adopted but then quickly abandoned), I prefer to embrace the quantity with greatest appeal to physical intuition–energy. To be sure, energy–especially energy flow with its degradation to lower temperatures, thus resulting in entropy production – is a more useful metric for quantifying complexity writ large. Not that energy has been overlooked in previous studies of Nature's many varied structures. Numerous researchers have championed energy's organizational abilities, including, for example, physicists (Morrison 1964 and Dyson 1979), biologists (Lotka 1922 and Morowitz 1968), and ecologists (Odum 1988 and Smil 1999).

Physical systems have always been well modeled by their energy budgets. But so are biological systems, now that science has abandoned the *élan vital* or peculiar "life force" that once plagued biology. Cultural systems, too, can be so modeled, for machines, cities, economies and the like are all described, at least in part, by energy flow. And it is non-equilibrium thermodynamics of open, complex systems that best characterizes resources flowing in and wastes flowing out, all the while system entropy actually decreases locally while obeying thermodynamics' cherished second law that demands environmental entropy increase globally.

Yet the quantity of choice cannot be simply energy, since the most primitive weed in the backyard is surely more complex than the most intricate nebula in the Milky Way. Yet stars have much more energy than any life form, and the larger galaxies still more. Our complexity metric cannot merely be energy, nor even just energy flow. That energy flow must be normalized to open systems' bulk makeup, enabling all such systems to be analyzed "on that same page." When this is done, as shown in Fig. 2.5, a clear and impressive trend is apparent – one of increasing energy per unit time per unit mass for a wide range of ordered systems throughout more than ten billion years of cosmic history.

Such an "energy rate density," Φ_m, is a useful way to characterize, indeed to quantify, complexity of a system–any system, physical, biological, or cultural (Chaisson 1998, 2001). This should not surprise us, since it was competing energy rate densities of radiation and matter that dictated events in the early Universe, as noted in the previous section.

Consider stars and their progressive changes. Stars do grow in complexity as their thermal and elemental gradients steepen with time; more data are needed to describe stars as they age. Normalized energy flows increase from protostars at "birth" (Φ_m ~0.5 erg/s/g), to main-sequence stars at "maturity" (~2), to red giants near "death" (~100). These values are essentially light-to-mass ratios, converting gravitational potential energy into luminosity rates and then normalizing by the mass of the system; the present-day Sun, for example, has 4×10^{33} erg/s and 2×10^{33} g, whereas a typical red-giant star (with increased internally ordered thermal and elemental gradients) has an order-of-magnitude higher luminosity for the same mass, hence a larger value of Φ_m. On and on, nuclear cycles churn; build up, break down, change – a kind of stellar "evolution" minus any genes, inheritance, or overt function, for these are the value-added qualities of biological evolution that go well beyond the evolution of physical systems.

Consider plants and animals. With few exceptions, rising complexity is evident throughout biological evolution, especially if modeled by energy-flow diagnostics. Life forms process more energy per unit mass (Φ_m $\sim 10^{3-5}$

erg/s/g) than does any star, and increasingly so with biological evolution. These values are specific metabolic rates, again normalizing incoming energy to system mass: plants, for example, need 17 kJ for each gram of photosynthesizing biomass and they get it from the Sun (only 0.1% of whose radiant energy reaches the planet's surface), thus for a biosphere of 10^{18} g, $\Phi_m \sim 10^3$ erg/s/g; more ordered 70-kg humans take in typically 2800 kcal/day and thus have a considerably higher value of $\Phi_m \sim 10^4$ erg/s/g; in turn, for human brains with ~20 W/day for proper functioning and a ~1300 g cranium, Φ_m is yet higher, $\sim 10^5$ erg/s/g (see Chaisson 2001 for many more such calculations). Onward across the bush of life – cells, tissues, organs, organisms – we find much the same story. Starting with life's precursor molecules and proceeding all the way up to plants, animals, and brains, the same *general* trend typifies life forms as for inanimate galaxies, stars or planets: The greater the perceived complexity of a system, the greater the flow of energy density through that system–either to build it or to maintain it, and often both.

Consider society and its cultural evolution. Once again, we can trace social progress in terms of normalized energy consumption for a variety of human-related advances among our hominid ancestors. Quantitatively, that same energy rate density increases from hunter-gatherers of a million years ago ($\Phi_m \sim 10^4$ erg/s/g), to agriculturists of several thousand years ago ($\sim 10^5$), to industrialists of contemporary times ($\sim 10^6$). Again, a whole host of energy per unit mass values can be used to track ancestral evolution, a highly averaged value of which today derives from 6 billion inhabitants needing ~18 TW of energy to keep our technological culture fueled and operating, thus Φ_m nearing 10^6 erg/s/g, and sometimes exceeding that for specialized energy needs (again, see Chaisson 2001, for a whole host of examples, many of which are plotted in Fig. 2.5). And here, along the path to civilization, as well as among the bricks, machines, and chips we've built, energy is a principal driver. Energy rate density continues rising with the increasing complexity of today's gadget-rich society.

Energy – the core of modern, non-equilibrium thermodynamics – ought not to be overlooked while seeking a broad, quantifiable metric for complexity. Whether acquired, stored, and expressed, energy has the advantage of being defined, intuitive, and measurable. Neither new science nor appeals to non-science are needed to justify the imposing hierarchy of our cosmic-evolutionary scenario, from stars to life to society.

Normalized energy flow also aids in unifying the sciences – namely, to diagnose aspects of physical, biological and cultural systems in a uniform manner, rather than fragmenting them further, indeed rather than complexifying unnecessarily the very subject of complexity science that we now seek to understand. More than any other single factor in science, energy flow would seem to be a principal means whereby all of Nature's ordered, diverse systems have naturally spawned rising complexity in an expanding Universe.

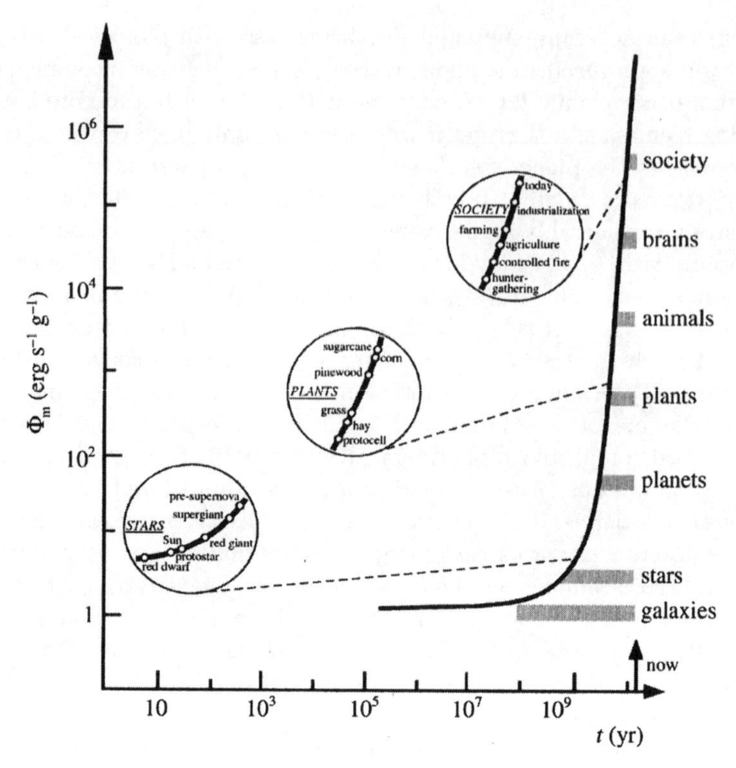

Fig. 2.5. The rise of free energy rate density, Φ_m, plotted as histograms starting at those times when various open structures emerged in Nature. Circled insets show greater detail of further measurements or calculations of Φ_m for three representative systems – stars, plants, and society – typifying physical, biological, and cultural evolution, respectively. Compare with the curve of rising complexity sensed from human intuition (Fig. 2.2) and that from our thermodynamic analysis of potentially increased order in a non-equilibrated cosmos (Fig. 2.4). To repeat, this is not to claim that galaxies per se evolved into stars, or stars into planets, or planets into life. Rather, this study suggests that galaxies produced environments suited for the birth of stars, that some stars spawned environments conducive to the formation of planets, and that at least one planet fostered an environment ripe for the origin of life – each system in turn able to handle increased amounts of energy flow per unit mass in an expanding Universe (Chaisson 2001)

References

Bonner JT (1988) Evolution of Complexity. Princeton University Press, Princeton, NJ.

Chaisson EJ (1998) Cosmic environment for the growth of complexity. BioSystems 46: 13.

Chaisson EJ (2001) Cosmic Evolution: Rise of Complexity in Nature. Harvard University Press, Cambridge, MA.

Dyson F (1979) Time without end: physics and biology in an open universe. Rev Mod Phys 51: 447.
Gold T (1962) The arrow of time. Am J Phys 30: 403.
Kauffman S (1993) Origins of Order. Oxford University Press, Oxford, UK.
Layzer D (1976) The arrow of time. Astrophys J 206: 559.
Lotka A (1922) Contribution to the energetics of evolution. Proc Nat Acad Sci 8: 147.
McMahon T and Bonner JT (1983) On Size and Life. WH Freeman, New York, NY.
Morowitz HJ (1968) Energy Flow in Biology. Academic Press, San Diego, CA.
Morrison P (1964) A thermodynamic characterization of self-reproduction. Rev Mod Phys 36: 517.
Odum H (1988) Self-organization, transformity, and information. Science 242: 1132.
Schroedinger E (1944) What is Life? Cambridge University Press, Cambridge, UK.
Smil V (1999) Energies: A Guide to the Biosphere and Civilization. MIT Press, Cambridge. MA.
Szathmary E and Smith JM (1995) Major evolutionary transitions. Nature 374: 227.

3 Stumbling into the MEP Racket: An Historical Perspective

Garth W. Paltridge

IASOS, University of Tasmania, GPO Box 252-77, Hobart, Tasmania, Australia

Summary. An historical tale is told of the author's involvement with research on the possible application of a principle of maximum entropy production to simulation of the Earth's climate system. The tale discusses a number of reasons why the principle took so long – and indeed is still taking so long – to become generally acceptable and reasonably respectable.

From my point of view the whole business of a principle of maximum rate of entropy production (MEP) emerged as a consequence of abysmal ignorance of much of the basic physics one is supposed to learn as an undergraduate. I never did understand thermodynamics – or indeed the purpose of it. It was taught to me in the old classic way as a never-ending stream of partial differentials which, at least at the time, didn't mean all that much. They meant even less when thrown together with a raft of pistons, cylinders and strangely behaving gases. As for things like entropy, enthalpy, Maxwell demons and Gibbs' functions, they were all quickly consigned to the scrap heap of memory as soon as the relevant course was finished. I suspect the words 'irreversible thermodynamics' never passed the lips of our lecturer. Mind you, I can remember being impressed with the Second Law, despite the fact no-one seemed quite to know what practical use it might be. The lack of an 'equals' sign anywhere in its exposition seemed to consign it to the realm of qualitative beauty rather than quantitative value. Had anyone ever made a dollar out of it?

So you may picture in the late 1960's a rather sub-standard physicist randomly tossed into the field of atmospheric physics and meteorology. He was basically an experimentalist, and thereby hoped to avoid displaying ignorance of the more esoteric and difficult areas of theoretical physics and applied mathematics. Perhaps 'randomly tossed' is putting it a bit high. In fact he actively chose the career because he had a vague feeling that running around in aeroplanes measuring things with weird instruments would be a lot of fun. And in this (about aeroplanes being fun) he was right. Where things went a bit pear shaped was when he discovered that atmospheric physics was, and still is, populated by extremely bright people working on some of the most fundamental problems of physics. To take just one example, one could refer to Von Karman who said somewhere – 'There are two great unexplained mysteries in our understanding of the universe. One is the nature of a unified generalised theory to explain both gravitation and electromagnetism. The other is an understanding of the nature of turbulence. After I die, I expect God to clarify the general field theory for me. I have no such hope for turbulence.' And as for pure meteorology, it turned out to be absolutely full of those wretched partial differentials, pistons, cylinders and strangely behaving gases.

Somewhere about that time also was the adolescence of the great new game of numerical modeling. It was (and of course still is) a gentlemanly activity, and many a scientist found himself or herself believing that numerical modeling was the *only* way to solve some of the great problems of the world. And after a while the exercise of pure simulation became an end in itself. The classic example was the modeling of climate, where it was necessary to introduce lots of tunable parameters so as to arrive at answers bearing at least some semblance of reality. The disease is still rampant today, although fairly well hidden and not much spoken of in polite society. The reader might try sometime asking a numerical climate modeler just how many tunable parameters there are in his latest model. He (the reader) will find there are apparently lots of reasons why such a question is ridiculous, or if not ridiculous then irrelevant, and if not irrelevant then unimportant. Certainly he will come away having been made to feel quite foolish and inadequate.

In fact the climate modeling business in the early seventies, although very impressive, did smack a little of describing the overall behaviour of a gas by simultaneously describing the motion of each and every molecule. There are after all some quite nice laws governing the macroscopic behaviour of a gaseous medium. So one could legitimately be rather arrogant and look down the nose on the subject and be rather nasty about it in public. Such an attitude was particularly attractive to someone for whom numerical modeling was another of the disciplines which fell (like thermodynamics?) into the too-hard basket. And it was during one of these looking-down-the-nose periods that the present author read somewhere that the last gasp of the physicist who couldn't solve a particular problem was to cast about for an extremum principle of some kind. What the reading didn't make clear was that any scientist worth his salt would at least have a feeling before he began what sort of extremum principle he was after.

In any event the teller-of-the-tale began a more-or-less random search for an extremum principle which might work with a simple one-and-a-half dimensional energy balance climate model. Putting that in English, he developed a model of the Earth's atmosphere and oceans in which adjacent boxes represented latitude zones (there were ten of them from pole to pole) and each box had a pair of separate sub-boxes which individually represented atmosphere and ocean as shown in Fig. 3.1. There were rather a lot of unknowns left over, even when he had cunningly used a number of tunable parameters to represent things like cloud albedo and cloud height and so on. The left-over unknowns boiled down to the surface temperature T, the cloud cover θ and the sum $LE{+}H$ of the surface-to-atmosphere latent and sensible heat fluxes of each box, together with the set X of north-south flows of energy between adjacent boxes. He had already woken up to the fact (obvious presumably to everyone else but new to him) that the real problem when trying to model climate is that the Almighty seems to have ensured that there are always more 'unknowns' than there are relevant equations. Funny that! As Von Karman implied, turbulence has a lot to answer for. Anyway, where an extremum principle might get into the act would be as a substitute for the missing relevant equations.

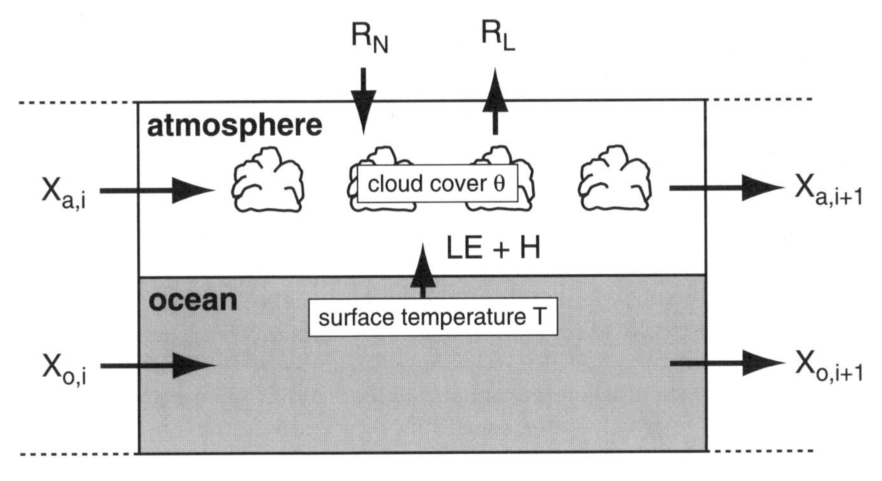

Fig. 3.1. Diagram of a latitude zone or 'box' of atmosphere and ocean with meridional energy fluxes X_o (in the ocean) and X_a (in the atmosphere) across latitudes i and $i+1$. The X of the text is the sum of X_o and X_a. The box has an ocean surface temperature T and an ocean-to-atmosphere non-radiant energy flux $LE+H$ of latent (LE) and sensible (H) heat. The fractional cloud cover of the box is θ. R_N and R_L are respectively the net short-wave and net long-wave radiation fluxes (at latitude i) into and out of the top of the box

It has to be admitted that the search involved a bit of cheating right at the beginning because there were only two energy balance equations which could be applied to each latitude zone – i.e., one at the top of the atmosphere and one at the ocean surface. The cheating took the form of a sort of subsidiary extremum principle. It was assumed that, given a particular net horizontal energy flux into a zone, its cloud cover and surface temperature would adopt values such that the vertical flux $LE+H$ from surface to atmosphere would be the maximum allowed by the two energy balance equations. There was some slight physical reasoning behind the assumption, but not so much that it would pass the censors. Suffice to say that the assumption gave good answers, so it didn't pay to be too critical.

Then it was simply a matter of looking at all sorts of strange overall parameters which might be made up from the individual variables calculated within the model. Among them were things like global-average surface temperature, average meridional flux, total solar radiation absorbed by the system and so on. In each case the distribution X of north-south energy flows between the boxes was juggled (this with a fancy numerical minimization routine) to see if the parameter had a minimum for a particular set Xp of the distribution X, and if so whether Xp and the associated cloud covers and surface temperatures of the zones looked anything like the real thing.

And so emerged a strange parameter involving the radiant fluxes into and out of the planet. Specifically, it was the sum over all the latitude zones i of the incoming net radiation ($R_N - R_L$ referring to the figure caption) divided by the outgoing infrared radiation R_L – that is, $\Sigma\{(R_{Ni} - R_{Li})/R_{Li}\}$. It worked beautifully. The only trouble was that, as a physical parameter, it didn't seem to mean much. In fact it didn't seem to mean anything at all, and eventually our intrepid investigator had to take the results to one of the old-style meteorologists who had a reputation for knowing what he was talking about. This was one Kevin Spillane, who immediately suggested taking the fourth root of the infrared radiation on the bottom line so that one would at least be dealing with recognizable units involving rate of energy flow divided by a temperature – that is, with units of the rate of entropy exchange. "So?" the author remembers saying. "What is entropy exchange and who cares?" Anyway, after something of a crash course on irreversible thermodynamics, he at last managed to convince himself that, if the results were to be believed, the atmosphere-ocean climate system seems to have adopted a format which *maximizes* the rate of entropy production within the system. The reader may note that it took some considerable time even to understand the reciprocal relation between entropy exchange and entropy production for steady state systems, and that minimization of the one was the equivalent of maximization of the other. To be fair, the physics behind the concept is not immediately obvious until one recognizes that the constraint of energy balance ensures comparison only of potential steady states of the system. The point is discussed again a little later in the paper. The overall entropy of any of these steady states must be constant, so in each case the internal rate of production must be balanced by the net rate of export across the boundary – i.e., out through the top of the atmosphere – via the radiative fluxes. The Second Law ensures that the internal entropy production is positive, so the net outward export is positive, and the net exchange (i.e., net inward flow) is negative because it is simply the outward export measured in the reverse direction. Mathematically, a minimum in the negative exchange has the greatest absolute value, and is the same as the maximum in the positive internal production.

Anyway, the result was ultimately published in a couple of papers (Paltridge 1975, 1978) in the Quarterly Journal of the Royal Meteorological Society. The second of them extended the idea a little, and among other things dealt with a 3-D '400-box' model which allowed calculation of the geographical distribution of cloud, surface temperature and horizontal energy flows in (separately) the atmosphere and the ocean. The journal referees of the time seemed to like the idea, and didn't give too much trouble.

And there matters stood for quite a large number of years. To be sure, a fair number of people addressed the issue in one way or another, and among other things confirmed the basic finding. They also provided a formal background to the analysis of entropy production associated with conversion of solar and thermal radiation from one 'temperature' to another. This was a considerable achievement, but as it turns out was probably fairly irrelevant to the particular issue of *why* the Earth-atmosphere system (or any other system for that matter) should adopt a format of maximum entropy production. Until that question could be answered, the MEP result could not be regarded, and rightly was not regarded, as anything other than a curiosity.

There were a number of things which didn't exactly help. Not the least of these was the rather forced and half-hearted physical explanation of the phenomenon which Paltridge himself propounded in a couple of associated papers (Paltridge 1979, 1981) in the late seventies and early eighties. It scarcely inspired confidence in the overall idea. But quickly setting that aside(!) some of the other unhelpful factors have at least an historical interest.

First, the seventies and early eighties were the great era of the sort of irreversible thermodynamics introduced by Prigogine and his colleagues. One of his theoretical results which had the simplicity to be well known and often quoted (though not perhaps really understood by a lot of people) was a principle of *minimum* entropy production. This was difficult to reconcile with a strange finding concerning maximum entropy production where, apart from anything else, the precise definition of entropy production was a bit loose. It required quite a lot of delving into the subject to appreciate that Prigogine's result applies to linear systems with fixed boundary conditions and (therefore) a single steady state. That single steady state is one of minimum entropy production relative to any non-steady condition to which the system might be pushed. The maximum entropy production concept concerns non-linear systems – so non-linear in fact that they can be thought of as having an infinite set of steady states, and by some magical means are able to select that particular steady state of their set which has the maximum production of entropy (see also Kleidon and Lorenz, this volume). The search for the 'magical means' was avoided by everyone.

Second, even if one can appreciate in principle the concept of a spectrum of potential steady states, it is not so easy to visualise a specific practical mechanism which has that peculiar characteristic. One is asking for a medium where the transfer coefficient (of the flux versus potential difference relation) can adopt any value it likes – a state of affairs which, even in principle, is difficult for any sensible fluid dynamicist to accept. The numerical modelers in particular are used to transfer coefficients which are proportional to some power of the potential difference, but such simple non-linear relations are still a long way from producing multiple possible steady states.

Third, there is no doubt that any result involving the word 'entropy' has a problem right from the beginning. For various rather obscure reasons, 'entropy' is a word that seems to attract the crackpots of the pseudo-scientific societies of the world. Its basic thermodynamic meaning is well enough defined, but its claim to universal application via the second law of thermodynamics is highly attractive to those who are, shall we say, rather more philosophic and hand-waving than is acceptable in the normal circles of the hard sciences. I have seen one of my early mentors pick out a madman in the audience of a scientific discussion simply because he (the madman) used the word 'entropy' in what might otherwise have been a quite sensible question. So one has to be a little careful not to be automatically assigned to the crackpot class when dealing with the subject. Perhaps this sort of thinking explains something of the fact that meteorologists and oceanographers and fluid dynamicists in general are far happier dealing with turbulent dissipation rather than the more general entropy production to which it is related.

And finally, when all is said and done, a global rather than a local constraint may be interesting physics but is not obviously useful in a world dominated by the numerical modeling of climate – that is, where the calculations done at each time step are inherently calculations about local conditions. One is apparently back to the problem with the second law itself – has anyone ever made a buck out of a global constraint?

Over the last little while the concept of maximum entropy production has got something of a new lease of life. More and more fluid-Earth (and indeed general planetary) examples have been proposed as cases where MEP might apply. The examples have provided hope, if not proof, that MEP might be used to bypass the difficulties of handling the specific processes of turbulence. Apropos of which, it is only over those last few years that it has been generally appreciated that the MEP principle, if it applies anywhere, must apply primarily to turbulent media where the necessary number and type of nonlinearities can pertain. Certainly, while in the earth-atmosphere context the dominant process of entropy production is associated with the downgrading of solar radiant energy to energy at terrestrial temperatures, that particular process (which is essentially linear) does not contribute directly to the creation of a set of potential steady states. Such a set derives specifically from the various *turbulent* transfer processes in the atmosphere and ocean.

Paltridge (2001) tried again to provide a physical explanation of why a turbulent medium might adopt the particular format associated with MEP. "Tried" is the operative word, since the explanation, while qualitatively acceptable (he supposes) as a physical picture – it is at least more acceptable than his earlier attempts 25 years before – still lacks the final touch of fully quantitative rigour. Basically the picture is of a turbulent medium transferring heat between two boundaries of different temperature maintained by an input of energy from outside the system. The system has an infinite set of possible steady states, each corresponding to a particular time-averaged distribution of the kinetic energy, eddy scale and physical position of the eddies in the medium, and each thereby corresponding to a particular value of transfer coefficient k. The set ranges from very large k (large heat transfer and, as a consequence, small temperature difference between the boundaries) to very small k (small heat transfer and, as a consequence, large temperature difference between the boundaries). The picture makes use of the fact that on short time-scales there are fluctuations of the instantaneous rate of heat transfer away from steady state because of the random hand-over of energy from one scale of eddy to another. There is a drift along the locus of steady states as the system returns towards a new steady state after each fluctuation. It turns out that the net drift due to random fluctuations is towards the middle of the set because the amplitudes of 'upward' and 'downward' fluctuations of heat transfer are different functions of the driving potential (i.e., of the temperature difference). Albeit with an assumption about the broad shapes of the fluctuation dependencies, it can be shown that the net drift is actually towards the steady state which has the maximum rate of thermodynamic dissipation or (and it is a slightly different steady state) towards the maximum rate of entropy production.

Among other things the explanation suggests the possibility that MEP might apply on a sufficiently local scale to be of use as a governing equation for the diffusive fluxes into and out of the grid boxes of the typical numerical climate model.

But the biggest fillip to the business has been Roderick Dewar's recent paper (Dewar 2003; also Dewar, this volume) which seems to provide what amounts to a statistical thermodynamic proof of the MEP concept. As I understand it (and lets face it I don't understand much of it yet – one's basic ignorance hasn't changed much in the last quarter of a century) Dewar has added what might be called a codicil to the second law of thermodynamics. Effectively he seems to have proved that, not only will an isolated system move ultimately to a state of maximum entropy as dictated by the second law, but it will get there as fast as it can. When his paper has been kicked around for a couple of years and is finally accepted by the gurus of theoretical physics, then perhaps we will at last have a basis for people to spend serious time finding applications for MEP. The numerical modelers might at last seize upon its respectability and do something with it (Ito and Kleidon, this volume; Shimokawa and Ozawa, this volume).

References

Dewar R (2003) Information theory explanation of the fluctuation theorem, maximum entropy production and self-organized criticality in non-equilibrium stationary states. J Phys A: Math Gen 36: 631–641.

Paltridge GW (1975) Global dynamics and climate – a system of minimum entropy exchange. Q J Roy Met Soc 101: 475–484.
Paltridge GW (1978) The steady-state format of global climate. Q J Roy Met Soc 104: 927–945.
Paltridge GW (1979) Climate and thermodynamic systems of maximum dissipation. Nature 279: 630–631.
Paltridge GW (1981) Thermodynamic dissipation and the global climate system. Q J Roy Met Soc 107: 531–547.
Paltridge GW (2001) A physical basis for a maximum of thermodynamic dissipation of the climate system. Q J Roy Met Soc 127: 305–313.

4 Maximum Entropy Production and Non-equilibrium Statistical Mechanics

Roderick C. Dewar

Unité d'Ecologie Fonctionelle et Physique de l'Environnement, INRA, BP 81, 33883 Villenave d'Ornon Cedex, France

Summary. Over the last 30 years empirical evidence in favour of the Maximum Entropy Production (MEP) principle for non-equilibrium systems has been accumulating from studies of phenomena as diverse as planetary climates, crystal growth morphology, bacterial metabolism and photosynthesis. And yet MEP is still regarded by many as nothing other than a curiosity, largely because a theoretical justification for it has been lacking. This chapter offers a non-mathematical overview of a recent statistical explanation of MEP stemming from the work of Boltzmann, Gibbs, Shannon and Jaynes. The aim here is to highlight the key physical ideas underlying MEP. For non-equilibrium systems that exchange energy and matter with their surroundings and on which various constraints are imposed (e.g., external forcings, conservation laws), it is shown that, among all the possible steady states compatible with the imposed constraints, Nature selects the MEP state because it is the most probable one, i.e., it is the macroscopic state that could be realised by more microscopic pathways than any other. That entropy production is the extremal quantity emerges here from the universal constraints of local energy and mass balance that apply to all systems, which may explain the apparent prevalence of MEP throughout physics and biology. The same physical ideas also explain self-organized criticality and a result concerning the probability of violations of the second law of thermodynamics (the Fluctuation Theorem), recently verified experimentally. In the light of these results, dissipative structures of high entropy production, which include living systems, can be viewed as highly probable phenomena. The prospects for applying these results to other types of non-equilibrium system, such as economies, are briefly outlined.

> *If one grants that [the principle of maximum Shannon entropy] represents a valid method of reasoning at all, one must grant that it gives us also the long-hoped-for general formalism for the treatment of irreversible processes ... [T]he issue is no longer one of mere philosophical preference for one viewpoint or another ; the issue is now one of definite mathematical fact. For the assertion just made can be put to the test by carrying out specific calculations, and will prove to be either right or wrong.*
>
> Jaynes (1979)

4.1 Introduction

Edwin Thompson Jaynes[1] (1922–1998) made many original and fundamental contributions to science in fields as diverse as applied classical electrodynamics, information theory, the foundations of probability theory, the interpretation of quantum mechanics, and radiation theory.

Much of his work is the expression of a single conviction, that probability theory – in which probability is interpreted in the original sense understood by Laplace and Bernoulli, as a measure of our state of knowledge about the real world – provides the uniquely valid rules of logic in science (Jaynes and Bretthorst 2003). In the vast majority of scientific problems actually encountered, we do not have sufficient information to apply deductive reasoning. What we need, said Jaynes, are the logic and tools of statistical inference (i.e., of probability theory) so that we may draw rational conclusions from the limited information we do have.

A key outcome of that conviction was Jaynes' reformulation of statistical mechanics in terms of information theory (Jaynes 1957a,b). This opened the way to the extension of the logic underlying equilibrium statistical mechanics (ESM) – implicit in the work of Boltzmann and Gibbs – to non-equilibrium statistical mechanics (NESM), as well as to many other problems of statistical inference (e.g., image reconstruction, spectral analysis, inverse problems). In all applications of this logic, the basic recipe consists of the maximisation of Shannon information entropy, subject to the constraints imposed by the available information – an algorithm now known as MAXENT (e.g., Jaynes 1985a).

How is it, then, that Jaynes' MAXENT formulation of NESM has for so long failed to be accepted by the majority of scientists when the logic of it is precisely that of Boltzmann and Gibbs?

Part of the answer lies with the relative paucity of published results from the MAXENT school (Dougherty 1994), especially with regard to new testable predictions far from equilibrium. The most extended account of Jaynes' NESM appears as part of a conference paper (Jaynes 1979, Sect. D). While that account makes clear the generality of the approach in principle, it is applied there within a perturbative approximation only to reproduce some known results for near-equilibrium behaviour.

Another reason why the MAXENT formulation of NESM has not caught on as it might have done almost certainly lies with the conceptual gulf between the Bayesian and frequency viewpoints of probability (Jaynes 1979, 1984). The frequency viewpoint – that probability is an inherent property of the real world (the sampling frequency) rather than a property of our state of knowledge about the real world (the Bayesian viewpoint) – dominated

[1] A biographical sketch and bibliography are available at http://bayes.wustl.edu/etj/etj.html

scientific thinking for much of the twentieth century. No wonder, then, that progress has been slow.

The hypothesis of maximum entropy production (MEP), which is explored by several authors in this volume, has likewise made slow progress. It has still to be widely accepted as a generic property of non-equilibrium systems despite a growing body of empirical evidence pointing in that direction (Lorenz 2003; Ozawa et al. 2003). The sticking point has been the perceived lack of a rigorous theoretical explanation for MEP.

So here we have, on the one hand, a theory in search of evidence (the MAXENT formulation of NESM) and, on the other hand, evidence in search of a theory (MEP). This Chapter gives an overview of some recent work proposing a mutually beneficial marriage between the two (Dewar 2003). As is often the case with such proposals, while the purpose might be well intentioned the result may be to have rocks thrown from both sides.

The main difficulties encountered at this stage are not so much technical as conceptual in nature, and so here I will try to give a non-mathematical account that emphasises the key physical ideas leading to MEP. I begin by briefly retracing the historical path of ideas from Boltzmann to Gibbs and Shannon which eventually led to Jaynes' MAXENT formulation of NESM. But if Jaynes' formulation is essentially an algorithm for statistical inference, what is the guarantee that it should work as a description of Nature? I discuss two key ideas of Jaynes – *macroscopic reproducibility* and *caliber* – that make the physical relevance of the algorithm intuitively clear (Jaynes 1980, 1985b).

Building on these ideas, I then present the path information formalism of NESM and discuss some new far-from-equilibrium predictions that have recently been obtained from it (Dewar 2003) – specifically, the emergence of MEP and self-organized criticality, and a result (known as the Fluctuation Theorem) concerning the probability of violations of the second law of thermodynamics.

I conclude that the MAXENT derivation of MEP explains its apparent prevalence throughout physics and biology, and suggests how MEP might be applicable to non-equilibrium systems more generally (e.g., economies, also see Ruth, this volume). In the light of this derivation, dissipative structures of high entropy production, which include living systems, may now be understood as phenomena of high probability.

4.2 Boltzmann, Gibbs, Shannon, Jaynes

Boltzmann interpreted Clausius' empirical entropy (S) as the logarithm of the number of ways (W), or microstates, by which a given macroscopic state can be realized ($S = k\log W$, where k is Boltzmann's constant). The second law of thermodynamics (maximum entropy) then simply means that the observed macrostate is the most probable one, i.e., it is the one that could be realized by Nature in more ways than any other. Microstate counting worked fine for isolated systems with fixed total energy and particle number.

Gibbs noted that Boltzmann's results could also be obtained by minimising the somewhat obscure quantity $\Sigma_i p_i \log p_i$ with respect to the microstate probabilities p_i, subject to the appropriate constraints on energy and particle number. Gibbs (1902) called the quantity $\Sigma_i p_i \log p_i$, or $< \log p_i >$, the 'average index of probability of phase'. He was then able to generalise ESM to open equilibrium systems, by extending the imposed constraints to include system interactions with external heat and particle reservoirs. However, just what the Gibbs algorithm meant, and its relation to Boltzmann's insight, remained obscure.

Much later, Shannon (1948) introduced the information entropy, $-\Sigma_i p_i \log p_i$, as a measure of the amount of missing information (i.e., uncertainty) associated with a probability distribution p_i. In the context of ESM, Shannon's information entropy measures our state of ignorance about the actual microstate the system is in. More quantitatively, a result called the Asymptotic Equipartition Theorem tells us that for systems with many degrees of freedom, the information entropy is equal to the logarithm of the number of microstates having non-zero probability (Jaynes 1979). It is a direct measure of the extent (or spread) of the distribution p_i over the set of microstates.

Jaynes' first insight was to see, in the light of Shannon's work, what the Gibbs algorithm meant and how it related to Boltzmann's insight. By maximising the information entropy with respect to p_i, Gibbs was constructing the microstate distribution with the largest extent compatible with the imposed constraints, thus generalising Boltzmann's logic of microstate counting to the microstate distribution (Jaynes 1957a,b).

But as Jaynes went on to realise, the Gibbs algorithm is much more than that. For the quantity $-\Sigma_i p_i \log p_i$ can be missing information about *anything*, not just about the microstates of equilibrium systems. Thus, during the late 1950s and early 1960s Jaynes developed his information theory formulation of NESM based on applying the Gibbs algorithm to non-equilibrium systems (Jaynes 1979).

But it did not stop there. Jaynes saw the Gibbs algorithm as a completely general recipe for statistical inference in the face of insufficient information (MAXENT), with useful applications throughout science, not just in statistical mechanics. Viewed as such, it is a recipe of the greatest rationality because it makes the least-biased assignment of probabilities, i.e., the one that incorporates only the available information (imposed constraints). To make any other assignment than the MAXENT distribution would be unwarranted because that would presume extra information one simply does not have, leading to biased conclusions.

4.3 Macroscopic Reproducibility

But if MAXENT is essentially an algorithm of statistical inference (albeit the most honest one), what guarantee is there that it should actually work as a description of Nature? The answer lies in the fact that we are only concerned with describing the reproducible phenomena of Nature.

Suppose certain external constraints act on a system. Examples include the solar radiation input at the top of Earth's atmosphere, the temperature gradient imposed across a Bénard convection cell, the velocity gradient imposed across a sheared fluid layer, or the flux of snow onto a mountain slope. If, every time these constraints are imposed, the same macroscopic behaviour is reproduced (atmospheric circulation, heat flow, shear turbulence, avalanche dynamics), then it must be the case that knowledge of those constraints (together with other relevant information such as conservation laws) is sufficient for theoretical prediction of the macroscopic result. All other information must be irrelevant for that purpose. It cannot be necessary to know the myriad of microscopic details that were not under experimental control and would not be the same under successive repetitions of the experiment (Jaynes 1985b). We can only imagine with horror the length of scientific papers that would be required for others to reproduce our results if this were not the case.

MAXENT acknowledges this fact by discarding the irrelevant information at the outset. By maximising the Shannon information entropy (i.e., missing information) with respect to p_i subject only to the imposed constraints, MAXENT ensures that only the information relevant to macroscopic prediction is encoded in the distribution p_i. Therefore, *if* we have correctly identified all the relevant constraints, then macroscopic predictions calculated as expectation values over the MAXENT distribution will match the experimental results reproduced under those constraints.

But of course that last *if* is crucial. In any given application of MAXENT there is no a priori guarantee that we have incorporated all the relevant constraints. But if we have not done so, then MAXENT will signal the fact *a posteriori* through a disagreement between predicted and observed behaviours, the nature of the disagreement indicating the nature of the missing constraints (e.g., new physics). MAXENT's failures are more informative than its successes. This is the logic of science.

Jaynes considered reproducibility – rather than disorder – to be the key idea behind the second law of thermodynamics (Jaynes 1963, 1965, 1988, 1989). Suppose that under given experimental conditions a system evolves reproducibly from initial macrostate A to final macrostate B (Fig. 4.1). The initial microstate lies somewhere in the phase volume W_A compatible with A, although we do not know where exactly because we cannot set up the system with microscopic precision. By Liouville's theorem, the system ends up somewhere in a new region of phase space W'_A having the same volume W_A. If the macroscopic transition A $\rightarrow$ B is reproducible for all initial microstates, then W'_A cannot be greater than the phase volume W_B compatible with B. Hence $S_A = k\log W_A = k\log W'_A \le k\log W_B = S_B$. This is the second law.

By the same token, while the reverse macroscopic process B → A (squeezing the toothpaste back into the tube) is possible because the microscopic equations of motion are reversible, it is not achievable reproducibly (i.e., it is highly improbable) because we cannot ensure by macroscopic means that the initial state lies in the appropriate subset of W_B (i.e., W'_A with all molecular velocities reversed) to get us back to A. Jaynes (1988) put some numbers to this: the probability of B → A, he conjectured[2], is something like $p = W_A/W_B = \exp(-(S_B - S_A)/k)$. If the entropy difference corresponds to just one microcalorie at room temperature, then we have $p < \exp(-10^{15})$. Although the macroscopic process B → A is possible, it is not macroscopically reproducible. The second law is the price paid for macroscopic reproducibility.

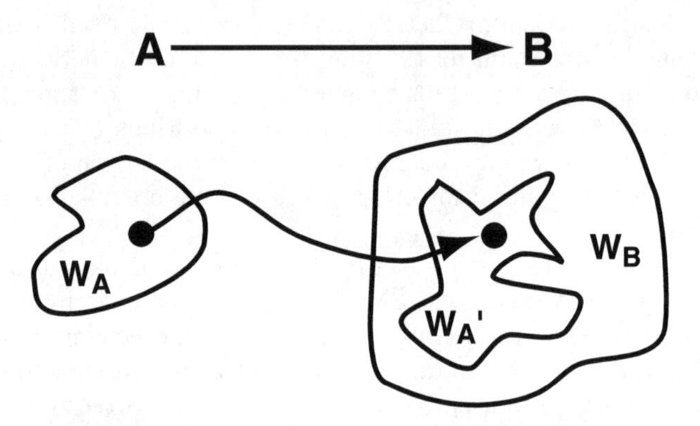

Fig. 4.1. The second law explained by macroscopic reproducibility (see text)

Already in 1867 James Clerk Maxwell understood that the second law was 'of the nature of a strong probability ... not an absolute certainty' like dynamical laws (Harman 1998). He introduced his 'finite being' (the term 'demon' was later coined by William Thomson) to underline this very point and to reject current attempts (notably by Clausius and Boltzmann) to reduce the second law to a theorem in dynamics. Perhaps Maxwell had something like Fig. 4.1 in mind when he wrote that 'the 2nd law of thermodynamics has the same degree of truth as the statement that if you throw a tumblerful of water into the sea you cannot get the same tumblerful out again' (Maxwell 1870). Without the means to identify and pick out the individual molecules involved, the process is effectively irreversible. Water flow is indeed a good analogue of Fig. 4.1; according to Liouville's theorem, probability in phase space behaves like an incompressible fluid.

[2] Jaynes's conjecture that 2nd law violating processes are exponentially improbable anticipates the Fluctuation Theorem (see below).

4.4 The Concept of Caliber

Mathematically, Jaynes expressed his MAXENT approach to NESM in a rather formal way, as the solution to a statistical inference problem of the most general kind (Jaynes 1979, Sect. D). From some known but otherwise arbitrary macroscopic history A, say, between times $t = -\tau$ and $t = 0$, we are asked to predict the macroscopic trajectory B for later times $t > 0$ (or to retrodict the previous history for $t < -\tau$). For non-equilibrium systems, A and B will generally involve both space- and time-dependent macroscopic quantities.

In Jaynes' application of MAXENT, one maximises the Shannon information entropy with respect to the microstate probability distribution $\rho(t = 0)$, subject to the known previous history A. In principle, one then integrates the (known) microscopic equations of motion to obtain $\rho(t)$ for all other times t, and constructs the macroscopic trajectory B by calculating the appropriate variables as expectation values over $\rho(t)$.

The maximised value of S depends on the known macroscopic history A, i.e., $S = S(A)$. Jaynes (1980) called that value the *caliber* of history A. It is a measure of the number of microstates at $t = 0$ compatible with that history. Equivalently, if you think of the previous history of each microstate for $-\tau < t < 0$ as a path in phase space, then the caliber is also a measure of the cross-sectional area of a tube formed by bundling together (like a stack of spaghetti) all those microscopic phase space paths compatible with the known macroscopic history A.

Jaynes (1980) suggested that the unknown macroscopic trajectory B could be inferred by an extension of the Gibbs algorithm that had given the caliber $S(A)$. Out of all possible macroscopic trajectories B, choose that one for which the combined caliber $S(A,B)$ is the greatest, because that is the one that could be realized by Nature in more ways than any other consistent with A. He later referred to this as the *maximum caliber principle*, and noted a tantalising analogy between the caliber in NESM and the Lagrangian in mechanics (Jaynes 1985b).

4.5 Path Information Formalism of NESM

Recently I reformulated Jaynes' MAXENT approach to NESM directly in terms of phase space paths (Dewar 2003). That is, one maximises the *path information entropy* $S = -\Sigma_\Gamma p_\Gamma \log p_\Gamma$ with respect to p_Γ, subject to the imposed constraints and other relevant information (e.g., conservation laws). Here p_Γ is the (Bayesian) probability of microscopic phase space path Γ, and the sum is over all paths permitted by the microscopic equations of motion. S is a measure of our state of ignorance about which microscopic path the system actually follows over time.

As an initial test, I applied this formalism to non-equilibrium stationary states in which all macroscopic variables are independent of time, although spatial variations (e.g., temperature gradients) are of course present. My main interest was to try to put maximum entropy production and a result known as the Fluctuation Theorem (Evans and Searles 2002) – which also concerns entropy production – on a common theoretical footing.

Why a path formalism? Because local conservation laws concern changes in energy and mass over time, their inclusion as fundamental microscopic constraints in MAXENT forces one to consider the behaviour of the system over time. Phase space paths rather than microstates are then the natural choice of microscopic description. This choice was also influenced by my reading of the literature on the Fluctuation Theorem, for which the standard proofs explicitly considered pairs of phase space trajectories related by path reversal (Evans and Searles 2002).

I also sensed that a path formalism was somehow truer to the spirit of Jaynes' maximum caliber principle, although at the time I did not see explicitly how his presentation of it (predicting unknown trajectory B from known history A) related to my steady-state problem[3].

Before discussing some specific predictions of this path information formalism of NESM, let us pause to state the problem more precisely and to recap the rationale for its solution by MAXENT, now in the twin contexts of reproducibility and maximum caliber.

Problem: Given the external constraints (and any other relevant constraints) acting on our system, which macroscopic steady state – among all possible macroscopic steady states compatible with those constraints – is the one selected by Nature? For example, given the input of solar radiation at the top of the Earth's atmosphere (and local energy and mass conservation), which climate state is selected?

Solution: The selected steady state is completely described by that path distribution which maximises the path information entropy S subject to those (and only those) constraints. Macroscopic quantities are calculated as expectation values over the MAXENT path distribution.

Rationale in terms of reproducibility: If Nature selects the state reproducibly, i.e., every time the constraints are imposed, then knowledge of those constraints alone must be sufficient to predict the result. Provided we have incorporated all the relevant constraints, macroscopic predictions inferred as expectation values over the MAXENT path distribution will agree with Nature's selected state.

Rationale in terms of maximum caliber: For systems with many degrees of freedom, the Asymptotic Equipartition Theorem implies that the value of S for a given path distribution p_Γ is the logarithm of the number of paths with non-zero probability. Let us call S the caliber of p_Γ (i.e., the path analogue of a microstate distribution's extent in phase space). The MAXENT path distribution is therefore spread across the largest number of paths compatible with the imposed constraints, i.e., it describes the macroscopic state that can be realized by more paths than any other.

---

[3] This only became clear to me during the preparation of the present article.

Reproducibility = maximum caliber: These two rationales are physically equivalent. The selected steady state is reproducible precisely because its macroscopic properties are characteristic of *each* of the overwhelming majority of possible microscopic paths compatible with the constraints. That is why it does not matter which particular microscopic path the system follows in any given repetition of the experiment.

In this formalism the caliber $S = -\Sigma_\Gamma p_\Gamma \log p_\Gamma$ is defined for any path distribution p_Γ, not just the MAXENT distribution. Using this more general definition of caliber (logarithm of the number of paths with $p_\Gamma > 0$), the maximum caliber principle is then identical to the Gibbs algorithm, rather than an extension of it (cf. Jaynes 1985b).

Finally it is worth emphasising that expectation values calculated from the path distribution are statistical inferences (Bayesian viewpoint). They are not the result of the system somehow sampling different paths in the real world (frequency viewpoint). While microstate distributions have always been open to an ergodic interpretation (expectation value = time average), clearly this interpretation makes no sense at all for the path distribution. In any given experiment the system only ever follows one microscopic path, S measures our state of ignorance about which one, and MAXENT predicts the macroscopic behaviour reproduced each time.

4.6 New Results Far from Equilibrium

The above formalism was used to predict the stationary steady-state 7properties of a general open, non-equilibrium system exchanging energy and matter with its external environment (Dewar 2003). The path information entropy $S = -\Sigma_\Gamma p_\Gamma \log p_\Gamma$ was maximised with respect to p_Γ subject to the relevant constraints, denoted collectively by A. Typically, these constraints consist of: ($A1$) Local energy and mass balance (conservation laws); ($A2$) Stationarity of the 'fast' system variables that are in an approximate steady-state on the timescale in question; ($A3$) External forcings (e.g., solar radiation input) by which the system is maintained out of equilibrium; ($A4$) Internal constraints (e.g., critical thresholds, genetic constraints, fixed system parameters) which, like $A2$, depend on the timescale in question: DNA, for example, is a fixed parameter over the lifetime of an organism, but a system variable on evolutionary timescales and thereby eventually becomes subject to selection by MAXENT. Note that $A1$ is a universal microscopic constraint common to all systems on all timescales, whereas the nature of constraints $A2$–$A4$ is specific to each system and timescale.

All macroscopic quantities can then be calculated as expectation values over the MAXENT path distribution. Which macroscopic quantities are we interested in? Typically these are the local distributions of heat and mass density within the system and the fluxes of heat and mass across the system boundary. These describe both the internal state of the system and its interaction with the external environment. Let us denote this macroscopic information collectively by B.

It then proves useful to apply the MAXENT algorithm in two steps, with the unknown macroscopic state B acting as a temporary constraint which is subsequently relaxed. S is first maximised with respect to the path distribution p_Γ, subject to constraint $A1$ and a trial state B. The maximised value of S after this first step (denoted S_1) depends on B, i.e., $S_1 = S_1(B)$. In the second step, $S_1(B)$ is maximised with respect to B, subject to the remaining constraints $A2$–$A4$. This completes the MAXENT algorithm. The choice of trial state B that maximises $S_1(B)$ represents the macroscopic steady state that is reproduced under the imposed constraints A. The final result for B is the same as if we had constructed the MAXENT path distribution imposed by A alone, and then calculated the properties of B directly as expectation values over that distribution.

We now note the close formal analogy between this procedure and Jaynes' procedure for inferring an unknown macroscopic trajectory B from a known macroscopic history A. Here we are inferring an unknown macroscopic steady state B from known constraints A. In each case the joint caliber $S(A,B)$ is maximised with respect to B, subject to A. In each case, if A is sufficient to reproduce B, then MAXENT will correctly predict the observed B.

4.6.1 Maximum Entropy Production (MEP)

After step 1 we find that $S_1(B) = \log W(EP{=}EP_{\mathrm{B}})$, where $W(EP{=}EP_{\mathrm{B}})$ is the number of paths whose thermodynamic entropy production rate (EP) equals that of macrostate B. The 'density of paths', W, is analogous to the density of states in equilibrium statistical mechanics. The thermodynamic entropy production rate that emerges here (EP) is just the familiar near-equilibrium expression involving products of fluxes and thermodynamic forces, but here it is also valid far from equilibrium. We have not assumed a local equilibrium hypothesis. The various contributions to EP derive directly from the fluxes (F) and sources (Q) in the local energy and mass balance (constraint $A1$), e.g., heat flow and frictional heating (from local heat balance), mass flow and chemical reactions (from local mass balance equations). The general rule giving the contributions to EP arising from each local balance equation is detailed below (Sect. 4.8).

In step 2 of the MAXENT algorithm we choose the value of B for which $\log W(EP{=}EP_{\mathrm{B}})$ is maximal, subject to $A2$–$A4$. This occurs when EP_{B} is maximal. Therefore step 2 is equivalent to MEP subject to $A2$–$A4$. The only requirement here is that the density of paths W is an increasing function of EP, so that a maximum in EP_{B} corresponds to a maximum in $\log W(EP{=}EP_{\mathrm{B}})$. That EP_{B} has a maximum reflects the trade-off between the component thermodynamic fluxes and forces, in which increased fluxes tend to dissipate the thermodynamic forces.

The conclusion here is that the MEP state is selected because it is the non-equilibrium steady state with the highest caliber, i.e., the one that can be realised by more microscopic paths than any other steady state compatible with the constraints. Because the thermodynamic entropy production emerges here directly from local energy and mass balance – the universal constraint $A1$ valid for all systems – it becomes clear why MEP is so prevalent across physics and biology. The predictions of MEP under constraints $A2$–$A4$ will vary from one system to another, and from one timescale to another, reflecting the specific nature of constraints $A2$–$A4$, but the principle of MEP itself would appear to have the same validity as local energy and mass conservation (constraint $A1$).

Provided we have identified all the relevant constraints, MEP will predict the experimentally reproduced result. Failure to do so will signal the presence of unaccounted constraints; but it could also indicate that we have ignored some contributions to the entropy production itself, signalling missing terms in our equations for local energy and mass balance.

4.6.2 The Fluctuation Theorem (FT)

Another general result that emerges from step 1 concerns the probability of violations of the second law (which, as Maxwell appreciated, is statistical in character).

Specifically, we find that the MAXENT probability of path Γ is proportional to $\exp(\tau EP_\Gamma/2k)$, where τ is the time duration of path Γ, EP_Γ is its entropy production rate, and k is Boltzmann's constant. Now consider the reverse path Γ_R obtained by starting from the end of path Γ and reversing all the molecular velocities so that we end up at the start of path Γ (i.e., reversing the curved path in Fig. 4.1). The entropy production rate of Γ_R is equal to $-EP_\Gamma$ by path-reversal symmetry of the microscopic equations of motion (i.e., all fluxes are reversed). This immediately implies that the ratio of the probability of Γ_R to that of Γ is equal to $\exp(-\tau EP_\Gamma/k)$.

As is easily shown, this result implies that the second law holds on the average, i.e.,$< EP > \geqslant 0$. It also says that second law violating paths with negative entropy production are possible, although exponentially improbable. This result is known as the Fluctuation Theorem (FT) (Evans and Searles 2002). The FT was first[4] derived heuristically in 1993. Subsequent derivations of the FT have been based on ergodicity and causality assumptions. Computer simulations of various models of microscopic dynamics have confirmed its validity. The first truly experimental verification of the FT was obtained in a delicate experiment which followed the Brownian motion of colloidal particles in an optical trap (Wang et al. 2002).

The path information formalism of NESM puts the FT and MEP on a common theoretical footing, and predicts that both are valid on very general grounds. The Bayesian rationale of MAXENT implies that ergodicity is not required to explain the FT. Rather, MAXENT suggests that the exponentially small probability of violations of the second law is, like MEP, charac-

[4] Jaynes (1988) was essentially there when he conjectured on grounds of macroscopic reproducibility (Jaynes 1963, 1965) that $p = W_A/W_B = \exp((-S_B - S_A)/k)$ for the entropy-consuming transition B $\rightarrow$ A (Fig. 4.1).

teristic of the reproducible behaviour of all systems obeying local energy and mass conservation.

4.6.3 Self-Organized Criticality (SOC)

Now we come to a result that emerged as an unexpected bonus (Dewar 2003). Some non-equilibrium systems such as earthquakes, snow avalanches, sand-piles, and forest fires tend to organize themselves into steady states which are characterised by large-scale fluctuations (Jensen 1998). This behaviour is reminiscent of equilibrium systems at phase transitions, obtained when variables such as temperature and pressure are tuned to critical values (e.g., the large-scale fluctuations in magnetisation produced when a ferromagnet is tuned to its Curie temperature). Only, many non-equilibrium systems appear to organize themselves into a critical state (SOC), apparently without tuning.

Nevertheless, all these systems are tuned to some extent. Typically they are forced out of equilibrium by a fixed but very slow input flux, F_{in} (of momentum, snow, sand, and lightning strikes, in the above examples – cf. constraint $A3$). In the archetypal example where a sprinkling of grains falls onto a sandpile (Bak et al. 1987), the slope of the sandpile tends to its largest possible value (the critical angle of repose), while critical fluctuations in the output grain flux about its average value (equal to F_{in} in the steady state) are induced in the form of sand avalanches of all sizes.

We can begin to understand SOC from an MEP perspective simply by noting that in the steady state, the sandpile entropy production is the product of the grain flux (F_{in}) and the slope. Because F_{in} is fixed, MEP predicts that the slope adopts its largest possible value. In other words, SOC is a special case of MEP applied to flux-driven systems. But what about the fluctuations?

In the path information formalism of NESM, fluctuations are described by the path distribution p_Γ. It can be shown from the path-reversal symmetry properties of p_Γ that, in the limit of slow input flux $F_{\text{in}} \to 0$, the variance of the magnitude of the output grain flux (the avalanches) is proportional to $1/F_{\text{in}}{}^2$ and therefore diverges to infinity as $F_{\text{in}} \to 0$, the characteristic signature of SOC. This result involves exactly the same mathematics as in classical theories of equilibrium phase transitions, with $F_{\text{in}}{}^2$ playing the role of the control parameter (cf. $|T - T_c|$, the amount by which a ferromagnet is cooled below its Curie temperature T_c) and the avalanche flux playing the role of the order parameter (cf. spontaneous magnetisation). As the control parameter is tuned to zero, the order parameter goes to zero but fluctuations in the order parameter emerge on all length scales (cf. divergence of magnetic susceptibility).

We can understand SOC intuitively from an information perspective. As $F_{\text{in}} \to 0$, the external constraint becomes scale-free. Provided there is no scale set by other internal constraints (e.g., friction), then the steady state reproduced under these constraints must also be scale-free. Consequently, the system is dominated by fluctuations on all scales.

4.7 Thermodynamics of Life

What conclusions may we now draw regarding the thermodynamics of life? Non-equilibrium dissipative structures, which include living systems, appear to be consistent with MEP. They couple extended regions of high order (e.g., convention cells, mass transport pathways) with localised regions of high dissipation (e.g., boundary layers, chemical reaction sites). The localised regions are responsible for most of the system entropy production, while the ordered regions act as transport structures which permit this entropy to be produced and exported at the greatest rate possible under the combined constraints of stationarity and local energy and mass balance. Far from equilibrium, the coexistence of ordered and dissipative regions produces and exports more entropy to the environment than a purely dissipative 'soup'.

Since Schrödinger's influential book *What is Life?* (Schrödinger 1944), discussion of the thermodynamics of life has taken a rather biocentric viewpoint along the lines that in order to maintain their internal order living systems must export entropy to their surroundings. This viewpoint sees life as constantly competing against the second law. But if we are to understand the emergence of living systems and other dissipative structures then it is the coexistence of ordered and dissipative regions that we need to focus on, and whose natural selection we need to explain. MAXENT provides the proper viewpoint – we take the imposed constraints as our starting point and we ask: which pattern of energy and mass flows is reproducibly selected under those constraints? In the light of the MAXENT derivation of MEP we can now view living systems (and dissipative structures more generally) as highly probable phenomena. They are selected because they are characteristic of each of the overwhelming majority of ways in which energy and matter could flow under the constraints imposed by local energy and mass conservation.

For me, some of the most exciting applications of MEP lie at the interface between biology and physics, from bioenergetics at the cellular level (see Juretić and Županović, this volume) to biosphere-climate interactions at the planetary scale (see Kleidon and Fraedrich, Toniazzo et al., this volume). Perhaps many aspects of biological function, which until now have been interpreted from an adaptive or evolutionary standpoint (e.g., leaf stomatal behaviour, plant architectural adaptations, or the evolutionary trend towards increased biodiversity), can be viewed from a new perspective, as manifestations of MEP. If that viewpoint proves a valid one, natural selection in both biology and physics could then be understood as expressions of the same basic concept, namely, selection of the most probable state.

4.8 Further Prospects

MAXENT is a general algorithm for predicting reproducible macroscopic phenomena under given constraints. The derivation of MEP from MAXENT suggests that MEP itself may apply beyond purely physical and biological systems involving energy and mass transfer. Specifically, if a system's macroscopic state B is described by a local variable ρ (e.g., the analogue of heat density) which obeys a balance equation $\partial\rho/\partial t = -\nabla \cdot F + Q$ (local rate of change = flux convergence + net local source), then the two-step MAXENT procedure will lead (after step 1) to the emergence of a generalised entropy production involving contributions from both flux F and source Q (e.g., analogues of thermal and frictional dissipation), and then (after step 2) to analogues of MEP, FT and SOC (Dewar 2003). Explicitly, the generalised entropy production takes the form $EP = \int_V \left(\bar{F} \cdot \nabla\theta + \theta\bar{Q}\right)$ where θ is the analogue of inverse temperature ($1/T$), the overbar indicates a time-average over interval τ, and the space integral extends over the system volume V.

For example, Jaynes (1991) anticipated the application of MAXENT to the prediction of macro-economic behaviour. Does MEP apply there? Are financial crashes SOC? Is there a 2nd law analogue for economies? A starting point would be to identify ρ, F and Q for economies, and to specify the micro- and macro-economic constraints that apply (cf. constraints $A1$–$A4$).

On the theoretical side, we can see the prospect of generalising the path information formalism of NESM to non-stationary macroscopic phenomena, in the spirit of Jaynes (1979). Can MEP be extended to time-dependent macroscopic trajectories such as cyclic steady states?

But above all, let us go ahead and apply the path information formalism of NESM and its predictions to as wide a range of real-world problems as possible. Then, as Edwin Jaynes would have put it, let the results speak for themselves.

References

Bak P, Tang C, Wiesenfeld K (1987) Self-organized criticality : an explanation of $1/f$ noise. Phys Rev Lett 59:381–384.

Dewar RC (2003) Information theory explanation of the fluctuation theorem, maximum entropy production and self-organized criticality in non-equilibrium stationary states. J Phys A 36:631–641.

Dougherty JP (1994) Foundations of non-equilibrium statistical mechanics. Phil Trans R Soc Lond A346:259–305.

Evans DJ, Searles DJ (2002) The fluctuation theorem. Adv Phys 51:1529–1585.

Gibbs JW (1902) Elementary principles of statistical mechanics. Reprinted by Ox Bow Press, Woodridge CT (1981).

Harman PM (1998) The natural philosophy of James Clerk Maxwell. Cambridge University Press, Cambridge UK.

Jaynes ET (1957a) Information theory and statistical mechanics. Phys Rev 106:620–630.

Jaynes ET (1957b) Information theory and statistical mechanics II. Phys Rev 108:171–190.

Jaynes ET (1963) Information theory and statistical mechanics. In: Ford KW (ed) Brandeis Summer Institute 1962, Statistical Physics. Benjamin, New York, pp 181–218.

Jaynes ET (1965) Gibbs vs Boltzmann entropies. Am J Phys 33:391–398.

Jaynes ET (1979) Where do we stand on maximum entropy? In: Levine RD, Tribus M (eds) The maximum entropy principle. MIT, Cambridge MA, pp 15–118.

Jaynes ET (1980) The minimum entropy production principle. Ann Rev Phys Chem 31:579–601.

Jaynes ET (1984) The intuitive inadequacy of classical statistics. Epistemologia 7(special issue):43–74.

Jaynes ET (1985a) Where do we go from here? In: CR Smith, WT Grandy (eds) Maximum entropy and Bayesian methods in inverse problems. D Reidel, Dordrecht, pp 21–58.

Jaynes ET (1985b) Macroscopic prediction. In: H Haken (ed) Complex systems – operational approaches in neurobiology. Springer-Verlag, Berlin, pp 254–269.

Jaynes ET (1988) The evolution of Carnot's principle. In: Erickson GJ, Smith CR (eds) Maximum-entropy and Bayesian methods in science. Kluwer, Dordrecht, vol. 1, pp 267–282.

Jaynes ET (1989) Clearing up mysteries – the original goal. In: Skilling J (ed) Maximum entropy and Bayesian methods. Kluwer, Dordrecht, pp 1–27.

Jaynes ET (1991) How should we use entropy in economics? Unpublished manuscript available at http://bayes.wustl.edu/etj/etj.html.

Jaynes ET, Bretthorst GL (ed) (2003) Probability theory : the logic of science. Cambridge University Press, Cambridge UK.

Jensen HJ (1998) Self-organized criticality. Cambridge University Press, Cambridge UK.

Lorenz RD (2003) Full steam ahead – probably. Science 299:837–838.

Maxwell JC (1870). Letter to John William Strutt. In: PM Harman (ed) The scientific letters and papers of James Clerk Maxwell. Cambridge University Press, Cambridge UK 2:582–583.

Ozawa H, Ohmura A, Lorenz RD, Pujol T (2003) The second law of thermodynamics and the global climate system – A review of the maximum entropy production principle, Rev Geophys 41: 1018.

Shannon CE (1948) A mathematical theory of communication. Bell Syst Tech J 27:379–423, 623–656.

Schrödinger E (1944) What is life? Cambridge University Press, Cambridge UK.

Wang GM, Sevick EM, Mittag E, Searles DJ, Evans DJ (2002) Experimental demonstration of violations of the second law of thermodynamics for small systems and short time scales. Phys Rev Lett 89, 050601.

5 Using Ecology to Quantify Organization in Fluid Flows

Robert E. Ulanowicz and Michael J. Zickel

University of Maryland Center for Environmental Science, Chesapeake Biological Laboratory, Solomons, MD 20688-0038 USA

Summary. Numerous applications of variational principles derived from physical thermodynamics have been made to the description of development in living systems. While some have met with varying degrees of success, it appears none of the measures from classical thermodynamics adequately incorporates the roles of intrinsic system constraints into a robust description of biotic development. The flow network measure *ascendency*, therefore, has been formulated to express more explicitly the constraints immanent in ecosystem trophic exchanges. Ascendency has wide applicability and can be used as well to provide a measure of the overall degree of organization inherent in a purely physical flow field, such as rates of energy exchange. It can also be employed to pinpoint the bottlenecks that control the fluid flow field.

5.1 Introduction

The body of phenomenology known as thermodynamics derives almost entirely from observations on physical systems. It remains rich, however, in its implications for living systems. Of especial interest to biologists is the concept of entropy, and particularly the derivative variational principles of minimal and maximal entropy productions. For example, one encounters the Prigoginian notion of minimal entropy production applied to living systems (Zotin 1972). Conversely, the tendency towards maximal entropy production finds application in the physical realm (Paltridge 1975, 2001, this volume) as well as the biological (Swenson 1989; Kleidon and Fraedrich, this volume; Toniazzo et al., this volume).

The extrapolation from the physical realm to the biological is not without its difficulties, however. While physical constraints, such as conservation of energy and mass, clearly apply, there seems to be a tacit consensus that internal constraints play a proportionately larger role in biological behaviors than they do among physical processes. Some look for a way around this by reformatting the laws of thermodynamics in unitary fashion (Hatsopoulos and Keenan 1965; Kestin 1976). To capture biological directions, Schneider and Kay (1994) proposed a corollary to the unitary formulation, whereby living systems always act to degrade existing gradients in exergy (energy available for work) at the maximal rate possible (see also Schneider and Sagan 2004).

For yet others, such reformulations do not incorporate sufficiently the informational constraints inherent in the processes that support life. Thus, Kauffman (1995) calls for a "Fourth Law of Thermodynamics" to fill the void. The utility of variational principles or goal functions as providing direction for the development of living systems was the subject of a recent symposium (Mueller and Leupelt 1998). The emerging consensus was that no single principle or goal function seems capable of adequately explaining the life process at all scales. Rather, each principle serves in its turn as an "orientor" that helps to guide, but not fully determine, the unfolding of living systems (Bossel 1998).

5.2 Constraint Among Biotic Processes

These limitations and inadequacies notwithstanding, a more effective quantification of the constraints intrinsic to biological systems appears desireable. It was, after all, Schroedinger's emphasis upon what he called "negentropy" that invigorated the search for ways by which biological constraints can be encoded in matter and which culminated in the discovery of DNA. "Negentropy", however, has been a difficult notion to quantify, and the limitations inherent in entropy as a state variable have circumscribed its possible role in the description of biotic processes.

Bearing these difficulties in mind, Ulanowicz (1980, 1986) made the decision to play down somewhat the energetic aspects of biology in order to highlight the role that emerging constraints play in organic development. He sought to develop a phenomenology of biological constraint by attempting to quantify the hidden agencies that channel biotic transfers along certain pathways. He remained confident that biotic constraints could be quantified, even in the absence of explicit knowledge about their constituent mechanisms – just as in thermodynamics it is possible to measure state variables without any concrete knowledge about microscopic details.

The system of interest for Ulanowicz was the flow network that depicts the transfers of material or energy between all pairs of predators and prey. He denotes the transfer of material or energy from prey (or donor) i to predator (or receptor) j as T_{ij}, where i and j range over all components of an n-member ecosystem. The total activity of the system is taken to be simply the sum of all system processes, $T_{..} = \sum_{i,j} T_{ij}$, or what is called the "total system throughput" (A dot as a subscript is taken to mean summation over that particular index).

The constraints inherent in the flow network are assumed to arise in connection with the increase in the influence of autocatalytic feedbacks as the ecosystem develops (Ulanowicz 1986). Such unspecified constraints serve to channel flow ever more narrowly along those pathways that most effectively participate in the autocatalytic processes. Alternatively, constraints may be regarded as anything that causes certain flow events to occur more frequently than others. With frequency thus in mind, one supposes that constraint is somehow connected with the joint probability that a quantum of medium is *constrained* both to leave i and enter j. This probability may be estimated by the frequency $(T_{ij}/T_{..})$. One then notes that the *less constrained* probability that a quantum merely leaves i for an unspecified destination can be acquired by summing the joint probability over all possible destinations. Such frequency becomes $(T_{i.}/T_{..})$. Similarly, the unconstrained probability that a quantum enters j is estimated by $(T_{.j}/T_{..})$. Finally, one reckons the probability that a quantum could make its way by pure chance from i to j, *without any constraint*, as the product of the latter two frequencies, or $(T_{i.}T_{.j}/T_{..}^2)$.

When Tribus and McIrvine (1971) defined information as "anything that causes a change in probability assignment", they essentially were equating information with constraint. Information theory, then, could provide the format for how one might quantify constraint. Strangely, however, information theory does not address information (constraint) directly. Rather it starts with a measure of the rareness of an event, as first postulated by Boltzmann (1872) to be $-k \log p$, where p is the normalized probability $(0 \leq p \leq 1)$ of the given event happening, and k is a scalar constant that imparts dimensions to the measure. One notices how for rare events $(p \approx 0)$, Boltzmann's measure is very large; whilst for very common events $(p \approx 1)$, it is vanishingly small.

Because the constraints that act to channel flows act to make certain things happen *more frequently* in a particular way, one expects that, on average, the probability of such constrained events would be greater than those of corresponding unconstrained events. The rarer (unconstrained or unguided) circumstance that a quantum leaves i and accidentally makes its way to j can be quantified by applying the Bolzmann formula to the last probability defined above, i.e., $-k \log(T_{i.}T_{.j}/T_{..}^2)$. The more frequent condition that a quantum is constrained both to leave i and enter j would give rise under Boltzmann's assumption to $-k \log(T_{ij}/T_{..})$. Subtracting the latter quantity from the former and combining the logarithms yields a measure of the information inherent in the hidden constraints that channel the flow from i to j, i.e., $k \log(T_{ij}T_{..}/T_{i.}T_{.j})$.

Finally, to quantify the average constraint at work in the system as a whole, one weights each such pair-wise measure by the corresponding joint probability of constrained flow from i to j and then sums over all combinations of i and j (Abramson 1963). That is,

$$AMC = k \sum_{i,j} \left(\frac{T_{ij}}{T_{..}} \right) \log \left(\frac{T_{ij}T_{..}}{T_{i.}T_{.j}} \right) \tag{5.1}$$

where AMC is the "average mutual constraint" (known in information theory as the average mutual information.)

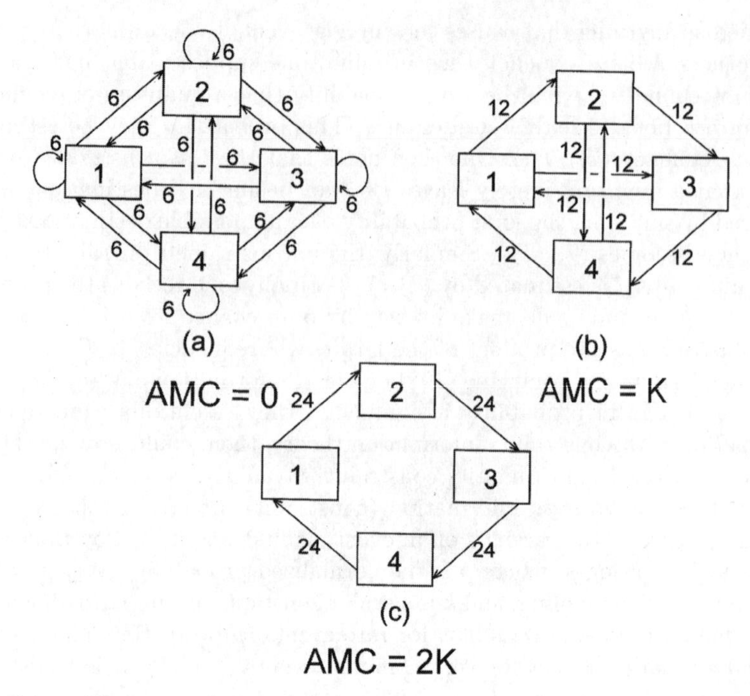

Fig. 5.1. **a** The most equivocal distribution of 96 units of transfer among four system components. **b** A more constrained distribution of the same total flow. **c** The maximally constrained pattern of 96 units of transfer involving all four components

To illustrate how an increase in *AMC* actually tracks augmented constraint, the reader is referred to the three hypothetical configurations shown in Fig. 5.1. In configuration (a) where medium from any one compartment will next flow is maximally indeterminate. *AMC* is identically zero. The possibilities in network (b) are somewhat more constrained. Flow exiting any compartment can proceed to only two other compartments, and the *AMC* rises accordingly. Finally, flow in schema (c) is maximally constrained, and the *AMC* assumes its maximal value for a network of dimension 4.

One notes in the formula for *AMC* that the scalar constant, k, has been retained. Tribus and McIrvine (1971) suggested that k be used to impart physical dimensions to an otherwise dimensionless information measure. Accordingly, the measure of constraint can be scaled by the total activity of exchange ($T_{..}$) to yield a "quasi-power" function called the system *ascendency* A, where

$$A = \sum_{i,j} T_{ij} \log \left(\frac{T_{ij} T_{..}}{T_{i.} T_{.j}} \right) \tag{5.2}$$

In his seminal paper, "The strategy of ecosystem development", Eugene Odum (1969) identified 24 attributes that characterize more mature ecosystems. These can be grouped into categories labeled species richness, dietary specificity, recycling and containment. All other things being equal, a rise in any of these four attributes also serves to augment the system ascendency (Ulanowicz 1986). It follows as a phenomenological principle that *"in the absence of major perturbations, ecosystems have a propensity to increase in ascendency."*

5.3 Quantifying Constraint in Fluid Flow

It is well and good that ecologists now have at their disposal a convenient measure of the level of constraint inherent in an ecosystem, seeing as how constraint appears to be a prominent aspect of living systems that heretofore had been insufficiently incorporated into conventional thermodynamic measures. The question of greater interest to the reader, however, is what relevance, if any, does this measure have to the disciplines of fluid flow, meteorology and climatology? (In the event a connection can be made, it would constitute an unusual "man bites dog" example of a concept first developed in the biotic sciences and then applied to the purely physical realm.)

To demonstrate the utility of ascendency to fluid mechanics, one begins with an arbitrary flow field of interest that is finite, continuous and can be divided into a countable number of finite elements that cover the field entirely and are contiguous with each other. Without loss of generality, it may be assumed that the flow field is rectangular and is divided by a rectilinear grid. The flow field can be one, two or three dimensional, for it is easily demonstrated that only a single index is necessary to uniquely identify any element in any finite spatial domain. For example, if the flow field is two-dimensional, one may divide the field into m rows of n cells each and number the cells consecutively 1,2,3,... m, m+1, m+2,..., $(mn-2), (mn-1), mn$. A similar scheme can be used to enumerate a three-dimensional field. Again, without loss of generality, further consideration will be limited to a two-dimensional $(m \times n)$ flow field.

It is assumed that a quantitative description of a fluid flow field, $v(x, y)$, can be provided either by some analytical means or a numerical process, and the values of $v(x, y)$ at any location (x, y) (and at the boundaries) are available with sufficient precision. (The dynamical case, $v(x, y, t)$ is considered below.)

Under these premises, the translation of the physical flow field $v(x, y)$ into an abstract flow network of dimensions $(mn \times mn)$ becomes straightforward. One begins by defining f_{ij} to be the total amount of fluid that passes from cell i to cell j during a unit of time. Only positive flows will be considered; that is, if a flow from i to j is calculated to be negative, then the absolute magnitude of the transfer is added to f_{ji}, instead of to f_{ij}.

Attention is now focused upon an arbitrary element k within the flow field. It exchanges fluid with elements $(k-1)$ and $(k+1)$ in the horizontal direction and with $(k-n)$ and $(k+n)$ in the vertical. For the moment attention is further narrowed upon the vertical line that separates spatial element k from element $(k-1)$. The amount of fluid passing this interface can be calculated as $\int_{k,(k-1)} v_x dy$, where v_x is the horizontal component of the velocity, and $k,(k-1)$ denotes a line integral over the vertical segment in question. Whenever this integral is positive, the calculated amount is added to the network element $f_{(k-1),k}$. If it is negative, the amount is added to $f_{k,(k-1)}$.

One can treat vertical transfers in a similar manner: Over the horizontal boundary separating k from $(k-n)$, one calculates the line integral $\int_{(k-n),k} v_y dx$. As before, if the result is positive, the magnitude is added to $f_{k,(k-n)}$; if it is negative, to $f_{(k-n),k}$.

By applying the first method to the interface between k and $k+1$ and the second to that separating k from $k+n$, one accounts for all exchanges involving element k. Obviously, one wishes to avoid any double counting of transfers, which can be accomplished by iterating over all internal *boundaries* (rather than the elements themselves), visiting each edge once and only once. Should the external boundary conditions happen to be impermeable, that is "no-flow", then the conversion to a network description of the fluid flow field is now complete. Whenever the boundary conditions are "wrap-around" (e.g., the right-hand side of element $2n$ is assumed to abut the left-hand side of element $[n+1]$), then the flows across these boundaries can be treated exactly like the internal boundaries. For more general boundary conditions, it will be necessary to increase the dimension of the flow matrix by at least one to $(mn+1)$ to be able to accommodate the external world. Accounting for boundary flows would then entail the calculation of elements like $f_{3n,(mn+1)}$ or $f_{(mn+1),(5n+1)}$, etc. The resulting flow matrix is likely to have high dimension and to be very sparse. (By "sparse" is meant that most matrix entries are zero.)

Having effected the conversion of a continuous (or approximately continuous) flow field into a discrete flow network, it is now but a formality to calculate the information indices that describe the status of ecosystem flow networks (Rutledge et al. 1976; Ulanowicz 1986; Ulanowicz and Norden 1990). As with the ecosystem trophic exchanges treated earlier, a dot is used as shorthand for summation over a subscript index.

The diversity of the flow field H can be defined using the familiar Shannon formula as

$$H = -\sum_{i,j} \left(\frac{f_{ij}}{f_{..}}\right) \log \left(\frac{f_{ij}}{f_{..}}\right) \tag{5.3}$$

This diversity, or complexity, encompasses both structured (constrained) and stochastic elements. Using Bayesian information theory, it becomes possible to parse out exactly how much of the calculated diversity can be characterized as structured from that which remains stochastic. As developed in the previous section, the amount of H which constitutes coherent (constrained) flow structure is assessed by the average mutual constraint as

$$AMC = \sum_{i,j} \left(\frac{f_{ij}}{f_{..}}\right) \log \left(\frac{f_{ij}f_{..}}{f_{i.}f_{.j}}\right) \tag{5.4}$$

That is, AMC becomes an index of the organization inherent in the flow field. Presumably, the AMC will also corrolate strongly with one or more of the scalar metrics pertaining to the fluid flow correlation tensor.

The amount of H that does not appear as structured flow, $(H - AMC)$, represents the residual incoherency Φ:

$$\Phi = -\sum_{i,j} \left(\frac{f_{ij}}{f_{..}}\right) \log \left(\frac{f_{ij}^2}{f_{i.}f_{.j}}\right) \tag{5.5}$$

That is, Φ should be an index of how stochastic or turbulent the flow field appears under the network representation. One notes that $H \geq 0$, $AMC \geq 0$, and $\Phi \geq 0$.

The working hypothesis now being investigated by the authors is that whenever a flow field undergoes a transition from laminar (highly organized) to turbulent flow, AMC will decrease dramatically and Φ will abruptly increase. Conversely, if an organized flow suddenly displaces a stochastic one (as in the sudden appearance of Bernard or Langumir cells), AMC should rise abruptly and Φ should fall correspondingly. A related example of how AMC can be applied to a field of migratory animals is provided in Ulanowicz (2000), who showed, for example, how the ascendency of a uniform rectilinear migration field increased when a barrier was introduced into the middle of the migrating animals. He also demonstrated how the ascendency of a field of random migrations was negligible in comparison with one where migrations were directed and distinct. Such differences almost certainly will appear in analogous fluid flow fields.

The conversion of dynamical flow fields, $v(x, y, t)$, into three-dimensional flow networks is rather straightforward: Instead of considering the four lines bounding the square grid, one treats the six sides of the cube that envelops k. It remains, then, only to define the expanded information measures that can be invoked to quantify the resulting 3-D network. As before, one defines f_{ijk} as the transfer from spatial element i to neighboring element j during time interval k. Again, the dot shorthand for index summation is employed. Pahl-Wostl (1995) showed how several coherencies are aggregated within the measure I_t which she calls the temporal information:

$$I_t = \sum_{i,j,k} f_{ijk} \log \left(\frac{f_{ijk}^2 f_{...}}{f_{ij.}f_{i.k}f_{.jk}}\right) \tag{5.6}$$

This index I_t can be decomposed into several components, each of which quantifies a different aspect of coherency, such as when a system begins to oscillate in response to a frequency in an imposed forcing function (Ulanowicz 1991).

5.4 Identifying Flow Bottlenecks

Because the information measures just introduced appear to parallel the metrics associated with the conventional correlation tensor, one might understand why the reader might want to question whether another set of seemingly redundant measures is really necessary? It should be pointed out, therefore, that the information calculus affords some very convenient mathematical properties not shared by the more conventional measures. In particular, the information format allows for the immediate calculation of a field of sensitivity indicators.

For example, above it was shown how scaling the AMC by the total system throughput yields a function called the system "ascendency" A:

$$A = f_{..}AMC \tag{5.7}$$

or

$$A = \sum_{i,j} (f_{ij}) \log \left(\frac{f_{ij} f_{..}}{f_{i.} f_{.j}} \right) \tag{5.8}$$

It happens that the ascendency as it appears in (5.8) is homogeneous in f_{ij}, so that one can immediately write the sensitivity of the ascendency with respect to any arbitrary flow, say f_{pq} as

$$\frac{\partial A}{\partial f_{pq}} = \log \left(\frac{f_{pq} f_{..}}{f_{p.} f_{.q}} \right) \tag{5.9}$$

One can then search this matrix of sensitivities for local maxima, which should indicate "hotspots' where the flow field as a whole is most sensitive to the particular transfer in question.

Ulanowicz and Baird (1999) used this formal scheme to appraise nutrient transfers in ecosystems. They had estimated parallel networks for the seasonal flows of carbon, nitrogen and phosphorus among the principal taxa of the Chesapeake ecosystem. Using those networks, they applied the sensitivity indices calculated from the last formula, to uncover the rate-limiting flows in the system. After the fact, they were able to demonstrate analytically that the maximal sensitivities indicated those elements that were rate-limiting in the sense of Justus von Liebig (1854). By analogy, it becomes possible to entertain the hypothesis that the maxima of the indicated sensitivities provide a convenient way of identifying the "bottlenecks" or control points in a fluid flow field.

5.5 Conclusion

One may hypothesize different levels of organization at the microscale, as characterized by different values of ascendency, should result in differing macroscopic states of the fluid flow field with contrasting rates of entropy production. Furthermore, the behavior of the ascendency index could provide additional insights about the organization of flow when MEP does not apply (e.g., smaller scales, departures from steady state). Using the ecological concept of ascendency could provide new and valuable contributions to the microscopic analysis of fluid flows and might also find fecund application to the related fields of meteorology and climatology.

References

Abramson N (1963) Information Theory and Coding. McGraw-Hill, New York, NY, 201p.

Boltzmann L (1872) Weitere Studien ueber das Waermegleichgewicht unter Gasmolekulen. Wien Ber 66: 275–370.

Bossel H (1998) Ecological orientors: Emergence of basic orientors in evolutionary self-organization. In: Mueller F, Leupelt M (eds) Eco Targets, Goal Functions, and Orientors. Springer-Verlag, Berlin, pp 19–33.

Hatsopoulos G, Keenan J (1965). Principles of General Thermodynamics. John Wiley, New York, NY.

Kauffman SA (1995) At Home in the Universe: The Search for Laws of Self-Organization and Complexity. Oxford University Press, Oxford, UK.

Kestin J (1976) The Second Law of Thermodynamics. Benchmark Papers on Energy Vol 5, Dowden, Hutchinson, and Ross, Stroudsburg, PA.

Liebig J (1854) Chemistry in its Application to Agriculture and Physiology. Taylor and Walton, London. 401 p.

Mueller F, Leupelt M (1998) Eco Targets, Goal Functions, and Orientors. Springer-Verlag, Berlin. 619p.

Odum EP (1969) The strategy of ecosystem development. Science 164: 262–270.

Pahl-Wostl C (1995) The Dynamic Nature of Ecosystems: Chaos and Order Entwined. John Wiley & Sons, New York, NY.

Paltridge GW (1975) Global dynamics and climate – a system of minimum entropy exchange. Q J Roy Met Soc 101:475–484.

Paltridge GW (2001) A physical basis for a maximum of thermodynamic dissipation of the climate system. Q J Roy Met Soc 127:305–313.

Rutledge RW, Basorre BL, Mulholland RJ (1976) Ecological stability: an information theory viewpoint. J theor Biol 57: 355–371.

Schneider ED, Kay JJ (1994) Life as a Manifestation of the Second Law of Thermodynamics. Mathematical and Computer Modelling 19: 25–48.

Schneider ED, Sagan D (2004) Into the cool. University of Chicago Press, Chicago, IL.

Swenson R (1989) Emergent attractors and the law of maximum entropy production: Foundations to a theory of general evolution. Systems Research 6: 187–197.

Tribus M, McIrvine EC (1971) Energy and information. Sci Am 225: 179–188.

Ulanowicz RE (1980) An hypothesis on the development of natural communities. J theor Biol 85: 223–245.

Ulanowicz RE (1986) Growth and Development: Ecosystems Phenomenology. Springer-Verlag, New York, NY, 203 p.

Ulanowicz RE (1991) Complexity: Toward quantifying its various manifestations. WESScomm 1:43–50.

Ulanowicz RE (2000). Quantifying constraints upon trophic and migratory transfers in spatially heterogeneous ecosystems. In: Sanderson J, Harris LD (eds) Series in Landscape Ecology: A Top-Down Approach. Lewis Publishers, Boca Raton, FL, pp. 113–142.

Ulanowicz RE, Norden J (1990) Symmetrical overhead in flow networks. Int J Systems Sci 21: 429–437.

Ulanowicz RE, Baird D (1999) Nutrient controls on ecosystem dynamics: The Chesapeake mesohaline community. J Mar Sci 19:159–172.

Zotin AI (1972) Thermodynamic aspects of developmental biology. Karger, Basel.

6 Cosmological and Biological Reproducibility: Limits on the Maximum Entropy Production Principle

Charles H. Lineweaver

School of Physics, University of New South Wales, Sydney, Australia

Summary. The Maximum Entropy Production principle (MEP) seems to be restricted to reproducible dissipative structures. To apply it to cosmology and biology, reproducibility needs to be quantified. If we could replay the tape of the universe, many of the same structures (planets, stars, galaxies) would be reproduced as the universe expanded and cooled, and to these the MEP principle should apply. Whether the concept of MEP can be applied to life depends on the reproducibility of biological evolution and therefore on our ability to distinguish the quirky from the generic features of life. Parallel long term experiments in bacterial evolution can be used to test for biological reproducibility.

6.1 Maximum Entropy Production and Reproducibility

The Maximum Entropy Production (MEP) principle suggests that structures that destroy gradients will arrange themselves such that a maximum amount of entropy is produced (within the given circumstances). On planets, MEP predicts that winds and currents driven by thermal gradients establish themselves in a way to maximize entropy production (Paltridge 1975, 1979; Lorenz et al. 2001; also several chapters in this volume).

In Boltzmann's derivation of the 2nd law of thermodynamics, the entropy was defined as: $S = k \log W$, where W is the number of microstates by which a given macrostate can be realized. We do not need to describe the microstates accurately and we do not need to know which one of them the system is in, but we do need to be able to count them. In computing the entropy, we are essentially quantifying our ignorance. The system could be hiding in W hiding spots – we do not know which one – so the larger W is, the larger our ignorance and the larger the entropy. In recent ground-breaking work, Dewar (2003) has provided a derivation of Maximum Entropy Production (also Dewar, this volume). Dewar points out that our ignorance can be interpreted as ignorance about anything, not just microstates, and therefore it can be applied to non-equilibrium systems (Jaynes 1957). In Dewar's derivation of MEP, the degrees of freedom are not the number of microstates W of equilibrium systems, available to a particle as in Boltzmann's derivation, but are paths available to the system. To make this conceptual shift we do not need equilibrium but we do need reproducibility, and thus reproducibility becomes the key aspects to whether MEP can be applied.

Reproducibility can be defined as follows. Let there be two macrostates A and B, each described by only a few parameters. If, each time we set up macrostate A under the same constraints and with the same values of the parameters, it evolves and arrives at B, we call this evolution reproducible. We would like to widen the range of applicability of MEP to cosmology and biology. However, since MEP is limited to reproducible dissipative structures we need to identify such structures in cosmology and biology.

6.1.1 Cosmological Reproducibility

Let us go back to a time 10^{-33} seconds after the big bang and watch another realization of the universe unfold. We will try to identify which structures are produced as they were in our Universe. Which macrostates are reproducible? We assume the same laws of physics, the same constants, forces and the same expansion. The universe begins again hot and dense, and as it expands it cools and rarifies just as it did the first time. As the temperature of the cosmic microwave background (T_{CMB}) falls below the rest masses of elementary particles and the binding energy of protons, neutrons, nuclei, atoms and molecules, these structures form like dew drops condensing out of cooling moist air (Fig. 6.1). Galaxies form again. Stars and planets condense from swirling dissipative accretion disks. Terrestrial planets form with iron cores and wet surfaces. Plate tectonics again slowly stirs and differentiates the crusts while thermal gradients stir up the oceans and atmospheres with currents, hurricanes and cumulonimbus clouds. These dissipative structures are the reproducible products of gravitational clumping and the thermal gradients it produces. We conclude that the MEP should apply to all of them.

6.1.2 The Entropy of an Observable Universe Must Start Low

The big bang model starts with matter and radiation in thermal and chemical equilibrium, and thus apparently the universe begins in a state of maximum entropy or heat death. However, if the universe starts in a state of maximum entropy, entropy cannot increase and any maximum entropy principle becomes an empty statement of initial conditions. Also, since life (and any other dissipative structure) needs gradients to form and survive, the initial condition of any universe that contains life will be one of low entropy, not high entropy. One cannot start an observable universe from a heat death.

The missing ingredient that solves this dilemma is gravity. Matter, evenly distributed throughout the universe, has much potential energy and low entropy. In the standard inflationary scenario describing the earliest moments after the big bang, matter originates from the decay of the evenly distributed potential energy of a scalar field during a short period at the end of inflation called reheating. 'False vacuum' decays into our true vacuum. Vacuum energy cannot clump. However, once the potential energy is dumped relatively uniformly into the universe in the form of relativistic particles, these can cool

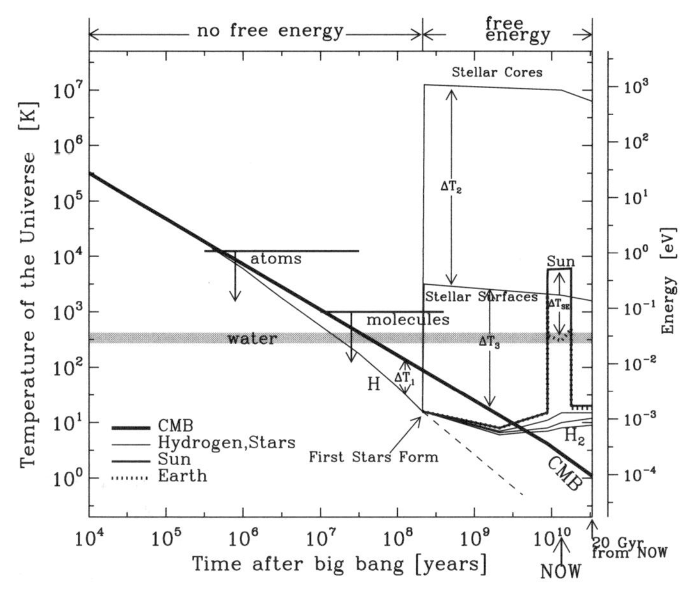

Fig. 6.1. Reproducible aspects of the evolution of our universe. As the universe expands, its temperature (the temperature of the cosmic microwave background "CMB") decreases as: $T_{CMB} \sim 1/\text{size}$. Half a million years after the big bang, the temperature of the universe falls beneath the binding energy of hydrogen. Atoms form. After the formation of neutral hydrogen, matter decouples from the CMB and the temperature of the matter decreases more rapidly than the CMB: $T_{matter} \sim 1/\text{size}^2$. For the first time in the history of the universe, matter and radiation are not in equilibrium with each other. This temperature difference is labeled 'ΔT_1' above. As the hydrogen cools further to $T \sim 20\,\text{K}$, clumps of it gravitationally collapse, heating up and reversing the thermal gradient between the CMB and hydrogen. Star formation begins about 180 million years after the big bang (Bennett et al. 2003). Balls of clumped hydrogen form stars that are $\sim 10^7\,\text{K}$ at their cores and $\sim 10^3 - 10^4\,\text{K}$ at their surfaces. This temperature difference is labeled 'ΔT_2' and is responsible for the convection cells on stellar surfaces as well as for complex stellar magnetic fields. The temperature difference between the surface of the stars and the CMB is labeled 'ΔT_3'. The Sun/Earth temperature difference responsible for all life on Earth is labeled $\Delta T_{SE.}$. The gravitational collapse and radioactivity inside the Earth set up a temperature difference between the center and the surface of the Earth of the same order of magnitude as ΔT_{SE}: 6000 K in the Earth's core and a surface temperature of $\sim 300\,\text{K}$. Thus the gravitational collapse of matter leads to thermal gradients, access to the free energy of nuclear fusion and to all the free energy driving terrestrial life. The current temperature of the CMB is 2.7 K. The energy scale in electron volts on the right helps make contact with the $\sim 0.2\,\text{eV}$ energy scales of the redox potentials that drove the molecular evolution that led to the origin of life (Nealson and Conrad 1999). For example, when ATP becomes ADP, 0.04 eV is released and photosynthesis extracts $\sim 1\,\text{eV}$ from each solar photon

and clump. The gravitational potential energy is enormous – analogous to a homogeneous distribution of boulders at all altitudes through the atmosphere. Thus, in this inflationary picture the potential energy of the vacuum is the ultimate source of all energy and the required low entropy initial state. The energy comes in the form of matter/antimatter pairs which annihilate and create a bath of photons. Because of an intrinsic asymmetry, the annihilation is incomplete and leaves one baryon for every billion photons. Their subsequent cooling (due to the expansion) and clumping of the baryons (due to gravity) is the source of all the free energy, dissipative structures and life in the universe.

6.1.3 Expansion Does Not Increase the Entropy of the Universe

In discussing maximum entropy production in the universe it is important to know what the entropy sources are, whether there is some maximum bound to the entropy of the universe (Fig. 6.2) and whether the expansion of the universe produces entropy.

It is difficult to talk about the total entropy in the universe without knowing how big the universe is. So we talk about the entropy in a representative sample of the universe. Typically we put an imaginary sphere around a few thousand galaxies and consider the entropy in this sphere. As the universe expands so does the sphere whose entropy we are considering. This is called the entropy per comoving volume. We parameterize the expansion of the universe with a scale factor R. This means that when the universe increases in size by a given factor, R increases by the same factor.

The entropy density s of a radiation field of temperature T is $s \sim T^3$. The entropy S in a given comoving volume V is $S = s\,V$. Since the comoving volume V increases as the universe expands, we have $V \sim R^3$. And since the temperature of the microwave background goes down as the universe expands: $T \sim 1/R$, we have the result that the entropy of a given comoving volume of space $S \sim R^{-3} * R^3 = \text{constant}$. Thus the expansion of the universe by itself is not responsible for any entropy increase. There is no heat exchange between different parts of the universe. The expansion is adiabatic and isentropic: $dS_{expansion} = 0$.

If expansion does not produce entropy, what does? Any region of the universe can be considered as an isolated cosmic box. The reason why entropy is increasing is because there are stars in that box. Hydrogen fuses to helium and nuclear energy is transformed into heat. Energy is released at the center of a star at millions of Kelvin and radiated away at thousands of Kelvin (ΔT_2 in Fig. 6.1). Dissipative stars extract energy at high temperature and discard it at low temperature.

To measure entropy in cosmology we just need to count photons. If the number of photons in a given volume of the universe is N, then the entropy of that volume is $S \sim k\,N$ where k is Boltzmann's constant. The vast majority of the entropy of the universe is in the cosmic microwave background. Stars cannot change that. If all the matter in the universe were transformed into 3 K blackbody radiation, the number of photons would add up to only $\sim 1\ \%$ of the number of CMB photons. The entropy of the universe would increase by only 1% .

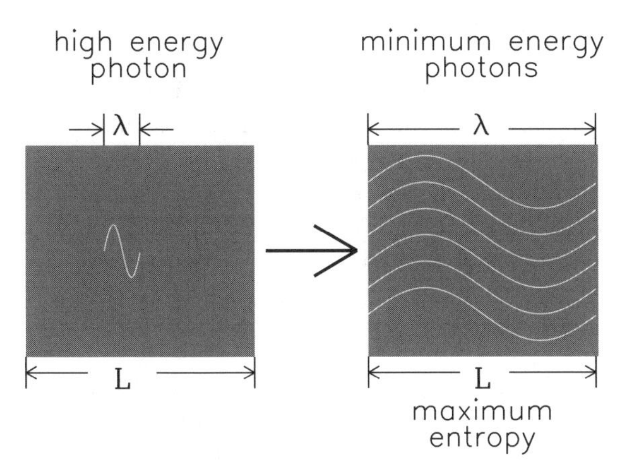

Fig. 6.2. The maximum entropy of the universe. The universe as a whole, or box-like partitions of it, can be treated as a closed system for which $dS \geq 0$. The maximum entropy of a closed system of size L is obtained when all the energy E, within the system is degraded into the smallest bits possible, i.e., all energy is converted into minimal energy photons with wavelengths as large as the system. This is the maximum entropy condition (Bekenstein 1981): $S_{max} = k\,N_{max} = k\,E/E_{min}$, where E is the energy within the comoving volume and the minimum quanta of energy is $E_{min} = hc/\lambda_{max} = hc/L$. Thus, we have $S_{max} = kEL/hc$, and the result is that the maximum entropy of the universe is proportional to the increasing size of the universe: $S_{max}(t) \sim L(t)$ (see Fig. 6.3 for limits on this size)

6.1.4 Return of the Heat Death

Before the discovery that 3/4 of the energy density of the universe was vacuum energy ($\Omega_\Lambda \sim 0.73$), it was thought that the expansion of the universe made the concept of classical heat death obsolete, because in an eternally expanding universe with an eternally decreasing T_{CMB}, thermodynamic equilibrium is a moving unobtainable target (e.g., Frautschi 1982). However, the presence of vacuum energy (also known as a cosmological constant) creates a cosmological event horizon (Fig. 6.3) and this imposes a lower limit to the temperature of the universe since the event horizon emits a blackbody spectrum of photons whose temperature is determined by the value of the cosmological constant:

$$T_\Lambda = 1/2\pi\,\Lambda^{1/2} \tag{6.1}$$

This is the minimum temperature that our universe will ever have if the cosmological constant is a true constant. Current values of Λ yield $T_\Lambda \sim 10^{-30}$ K. This new fixed temperature puts an upper bound on the maximum entropy of the universe and therefore reintroduces a classical heat death as the final state of the universe.

To summarize our cosmological considerations: Galaxies, stars and planets are reproducible structures and should be describable by MEP (see also Sommeria, this volume). The expansion of the universe by itself produces no entropy. Stars are currently the largest producers of entropy in the universe but all the stars in the universe will only ever be able to produce about 1% of the entropy contained in the CMB. The newly discovered cosmological constant limits the maximum entropy of the universe, and consequently the universe is on its way to a heat death.

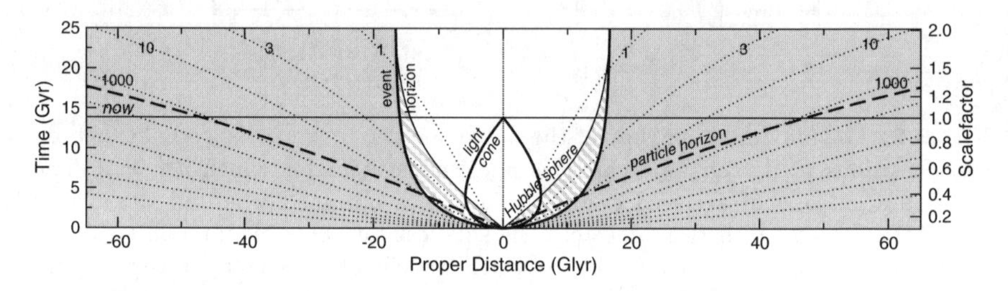

Fig. 6.3. The maximum size L of the system in Fig 6.2 is the cosmic event horizon shown here. As the universe expands the only part of it we can see is along our tear-drop shaped past light cone. As the universe gets older, our past light cone asymptotically approaches the event horizon. Our worldline is the central vertical line. Distant galaxies recede from us along the *dotted lines* – the worldlines of galaxies with currently observed redshifts of 1, 3 and 10 are labeled. Since the energy density of the universe is dominated by a cosmological constant Λ, the universe has an event horizon whose largest radius will be $\sim$ 18 billion light years ('Glyr'). Therefore the longest wavelength photon that will fit in the universe (a photon of the lowest possible energy) will have a wavelength that spans the universe: $\lambda_{max} \sim$ 36 billion light years. The cosmic event horizon imposes a maximum physical size to the observable universe and therefore a maximum wavelength of light λ_{max}. Therefore, since $S_{max}(t) \sim L(t) \rightarrow L_{max}$, S_{max} approaches a constant. The temperature of the universe approaches T_Λ and a heat death for the universe is possible ($S_{universe} \rightarrow S_{max}$). In such a situation the energy within the event horizon goes down and one would expect S_{max} to decrease. However, in Davis, Davies and Lineweaver (2003) we showed that the loss of entropy due to loss of energy is compensated exactly by the increasing entropy of the increasing area of the cosmic event horizon (figure from Davis and Lineweaver 2004)

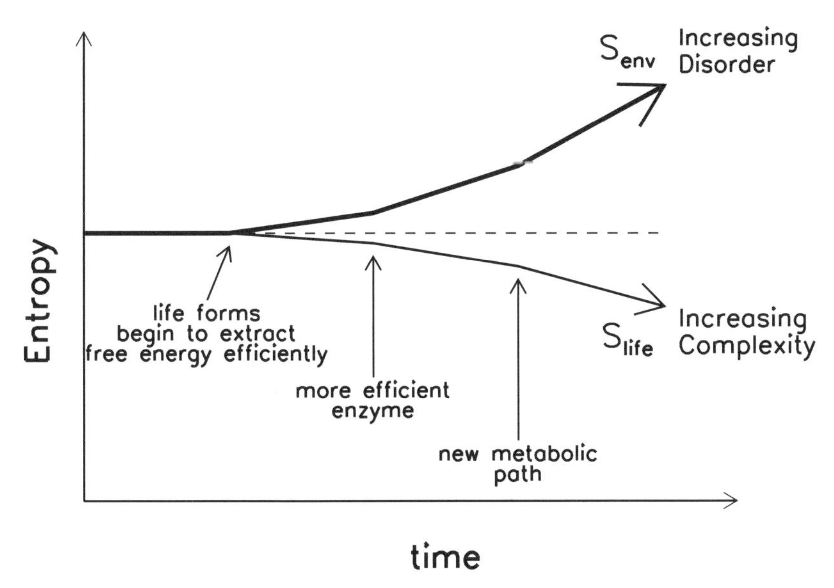

Fig. 6.4. Consider the entropy budget of a dissipative system and its immediate environment (S_{life} and S_{env} respectively). Dissipative systems have a low internal entropy maintained by the export of entropy to the environment: $dS = dS_{life} + dS_{env} \geq 0$. If the system is such that its order is increasing, $dS_{life} < 0$, this necessarily happens at the expense of the environment and we have $dS_{env} > |dS_{life}|$. The decreasing entropy of life does not violate the 2nd law since dS_{env} more than compensates for the lowering of entropy inside life. As life evolves and its metabolic paths become more efficient at extracting available free energy, this should lead to changing slopes as shown in the diagram (see also Chaisson, this volume)

6.2 Biological Reproducibility

6.2.1 Does Life Increase the Total Entropy Growth over What It Would Be Without Life?

Much evidence supports the idea that life increases the rate of entropy production (Fig. 6.4). For example, forests absorb more solar radiation by their lower albedo, and are cooler than deserts at the same latitude and thus produce more entropy. The decreasing S_{life} of Fig. 6.4 represents the increasing complexity of biological evolution. This trend is presumably due to the fact that life forms that can extract more work (and therefore produce more entropy) survive preferentially (Lotka 1922a,b; Ulanowicz and Hannon 1987). Ulanowicz and Hannon (1987) describe this as: "If two systems receive the same quantity of energy at the same entropy, that system which extracts the most work from its input before releasing it to its environment (as it inevitably must) can be said, in the second law sense of the word, to be the more efficient utilizer. Having extracted more work from the given amount of energy, the quality of the release is less, i.e., its entropy is higher." Thus, the evolution of more efficient metabolisms should be equivalent to the evolution of larger entropy production.

Figure 6.4 represents biological dissipative structures increasing net entropy ($dS = dS_{sys} + dS_{env} \geq 0$). We simplistically assumed no change in entropy due to the abiotic processes on the planet (horizontal dashed line). More realistically we need to include the entropy production by abiotic processes on a planet (Fig. 6.5).

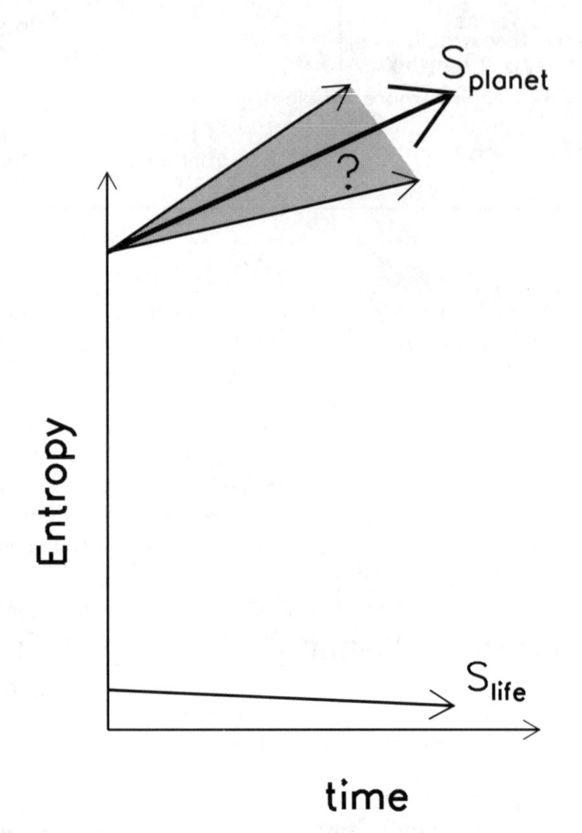

Fig. 6.5. Entropy produced by a planet (and life on that planet) as a function of time. The small decreasing entropy of life is negligible compared to the entropy and entropy production of the planet. If feedback mechanisms regulate the temperature of the planet, life can either increase or decrease the entropy production of the planet (grey region around S_{planet})

The conceptual Daisyworld model of Watson and Lovelock (1983) provides an example to investigate the role of biotic effects on planetary entropy production (see also Toniazzo et al., this volume). In Daisyworld, daisy albedo regulates planetary temperature. The fact that the Sun's luminosity increases on long time scales puts the focus on temperature reduction (= entropy increase). However, a symmetry is assumed, that it is just as easy to increase the temperature as it is to decrease it. Thus the assumption is made that entropy decrease is just as likely to occur as entropy increase. Such feedback mechanisms between dissipative systems are candidates for a violation of MEP if it can be shown that they arise reproducibly. It is still an open question whether such feedback mechanisms are symmetric with respect to entropy production (Ulanowicz and Hannon 1987). The issue is not whether life can make $dS_{planet} < 0$, but whether life can make dS_{planet} lower than it would be without life (under the second law constraint that $dS_{planet} > 0$). In Fig. 6.5, this symmetry and uncertainty are reflected by the symmetry of the grey area around S_{planet} and by the question mark in the case of entropy decrease. This issue is important for proposed resolutions to the faint early Sun paradox, which e.g., invoke biotic methanogenesis to warm the early Earth and reduce its entropy production (Pavlov et al., 2000).

6.3 Applying the Maximum Entropy Principle to Biological Evolution

One problem with applying MEP to life is the identification of the constraints (e.g., Lagrange multipliers). One by one, life can explore and reach out to influence the constraints and one by one the system can modify the previously "external" constraints. If this is happening continuously, then at any one time the current entropy maximum will be a local maximum not a global one, for it will be replaced by a larger maximum as soon as life figures out how to tap into other sources of free energy. This evolution is shown as the slope changes in Fig. 6.4. Since the number and complexity of constraints is large, this process can continue as long as untapped sources of free energy are available. Thus we hypothesize that MEP prescribes stable maxima for non-living dissipative structures and transient local maxima for life (see Kleidon and Fraedrich, this volume, for potential global maxima in entropy production at the planetary scale). In the general debate surrounding Lovelock's Gaia hypothesis (Lovelock 1972, 1988; Lovelock and Margulis 1974) a central issue is whether the biosphere, without other biospheres to compete with, can evolve in a way analogous to the way more traditional units of life, e.g., species, evolve (Lovelock 1988, Dawkins 1982). If competition and natural selection are the only drivers of evolution then the idea of Gaia evolving without competition seems inappropriate. However, if the second law of thermodynamics, in the form of maximum entropy production, can be successfully used to describe evolution, then considering the biosphere as an evolving life form seems more appropriate. In Dewar's derivation, competing micropaths lead to a global state of maximum entropy production. These micropaths, or degrees of freedom, do not have to be realized for the macroscopic steady-state to establish MEP. It also seems that MEP will be established globally independent of whether other Gaias are realized to compete with our Gaia. This contrasts with Dawkins' view that the competitors need to be present for selection to result in the evolution of Gaia.

6.4 Does the MEP Imply That Life Is Common in the Universe?

We would like to know how common life is in the universe. The rapidity of terrestrial biogenesis is sometimes invoked to support the idea that life is common in the universe (Lineweaver and Davis 2002). de Duve (1995) has argued from a biochemical point of view that life is a cosmic imperative. Does the MEP have anything to say on this issue?

Among the structures in the universe, far from equilibrium dissipative structures are ubiquitous and inhabit regions of thermal and chemical gradients (Prigogine 1980). Stars, convection cells, whirlpools, and hurricanes are common. MEP should apply to these reproducible macrostates. The formation of auto-catalytic reactions that live off of chemical gradients could be considered one of the earliest deterministic steps in the chain of molecular evolution that led to chemical life. Whether biogenesis is reproducible is unclear and without this MEP may not be applicable to biotic activity.

However, once we have biogenesis, can MEP be applied to photosynthesis or a given species? Surely there must be a spectrum of reproducibility between generic features that are reproduced (galaxies, stars and planets) and unique quirks that are not (tuataras, sulfur-crested cockatoos, HIV). One way to begin to determine this spectrum of reproducibility is by doing controlled experiments in evolution. Long term experiments in bacterial evolution can be used to test the reproducibility of metabolic adaptations to external stress such as temperature, pH and low food levels (Lenski 1998). In addition, if careful measurements of the entropy of the input nutrients and output waste can be made over long periods, evolution towards (or away from?) entropic maxima can be quantified.

References

Bekenstein JD (1981) Universal upper bound to entropy-to-energy ratio for bounded systems. Phys Rev D 23: 287–298.

Bennett CL et al. (2003) The Wilkinson Microwave Anisotropy Probe (WMAP) Basic Results. Astrophys J Supp Ser 148: 1–27.

Davis TM, Davies PCW, Lineweaver CH, (2003) Black hole versus cosmological horizon entropy. Classical and Quantum Gravity 20: 2753–2764. astro-ph/0305121.

Davis TM, Lineweaver CH (2003) Expanding Confusion. Pub Astron Soc Australia 21: 97–109. astro-ph/0310808.

Dawkins R (1982) The Extended Phenotype. Oxford University Press, Oxford, pp. 234–237.

de Duve C (1995) Vital Dust. Basic Books, New York.

Dewar RC (2003) Information theory explanation of the fluctuation theorem, maximum entropy production and self-organized criticality in non-equilibrium stationary states. J Phys A 36: 631. also at http://arxiv.org/abs/condsmat/0005382.

Frautschi S (1982) Entropy in an Expanding Universe. Science 217: 593–599.

Jaynes ET (1957) Information theory and statistical mechanics II. Phys Rev 108:171–190.

Lenski RE et al. (1998) Evolution of competitive fitness in experimental populations of E.coli: What makes one genotype a better competitor than another? Antonie van Leeuwenhoek 73: 35–47.

Lineweaver CH, Davis, TM (2002) Does the rapid appearance of life on Earth suggest that life is common in the Universe? Astrobiology 2/3: 293–304. astro-ph/0205014.

Lorenz RD, Lunine JI, Withers PG, McKay CP (2001) Titan, Mars and Earth: Entropy production by latitudinal heat transport. Geophys Res Lett 28, 415–418.

Lorenz R (2003) Full Steam Ahead-Probably. Science 299:837–838.

Lotka AJ (1922a) Contribution to the energetics of evolution. Proc Natl Acad Sci USA 8: 147–151.

Lotka AJ (1922b) Natural selection as a physical principle. Proc Nat Acad Sci USA 8: 151–154.

Lovelock JE (1972) Gaia: A new look at life on Earth. Oxford University Press, Oxford.

Lovelock JE (1988) Ages of Gaia. Norton:NY.

Lovelock JE, Margulis L (1974) Atmospheric homeostasis by and for the biosphere: the Gaia hypothesis. Tellus 26: 2–10.

Nealson KH, Conrad PG (1999) Life, past, present and future. Phil. Trans. Roy. Soc. Lond. B 354:1–17.

Paltridge GW (1975) Global dynamics and climate – a system of minimum entropy exchange, Q J R Meteorol Soc 101: 475–484.

Paltridge GW (1979), Climate and thermodynamic systems of maximum dissipation. Nature 279: 630–631.

Pavlov A, Kasting J, Brown L, Rages K. Freedman R (2000) Greenhouse warming by CH_4 in the atmosphere of early Earth. J Geophys Res 105:11981–11990.

Prigogine I (1980) From Being to Becoming. WH Freeman, New York.

Ulanowicz RE, Hannon BM (1987) Life and the Production of Entropy. Proc R Soc Lond B 232: 181–192.

Watson AJ, Lovelock JE (1983) Biological homeostasis of the global environment: the parable of Daisyworld. Tellus 35B: 284–289.

7 Entropy Production in Turbulent Mixing

Joël Sommeria

Laboratoire des Ecoulements Géophysiques et Industriels, LEGI/Coriolis, 21 Avenue des Martyrs, 38 000 Grenoble, France

Summary. We review the statement and application of a Maximum Entropy Production (MEP) principle for the modeling of turbulence. More specifically it applies to two-dimensional turbulence, for which a formalism of statistical mechanics has been proposed. In that case entropy measures the randomness of turbulent fluctuations rather than the molecular fluctuations considered in usual thermodynamics. Nevertheless the same MEP formulation also applies in usual thermodynamics, and we first show how it provides a general understanding of the classical diffusion law. In all cases the outcome of this MEP is a form of diffusion law, with additional terms taking into account long-range interactions (it provides generalized Fokker-Planck equations). This outcome is not unique, but depends on the assumed constraints. We show how the turbulence model can be improved by taking into account all the conservation laws, and by sorting out the deterministic effects of small eddies, limiting MEP to the unknown, random contribution. Finally the extension to other systems with long-range interactions is briefly discussed. This includes applications to gravitational systems. Remarkably, the same transport equations apply in biology, to the chemotactic aggregation of bacterial population.

7.1 Introduction

The "principle" of maximum entropy production (MEP) is a guideline to guess the evolution of a complex system. Although this idea is appealing, its scientific status is not clear. Different principles of maximum entropy production have been proposed in different contexts, as reviewed in this book, and it is not clear whether a unique physical law can emerge.

We discuss here how this idea of maximum entropy production can be specified and applied to the modeling of some macroscopic systems, ranging from turbulence to the dynamics of galaxies. Entropy measures the number of available microscopic states corresponding to some macroscopic state. The microscopic states generally represent thermal molecular motion. In our context, it will rather represent the turbulent fluctuations, while the macroscopic state represents some mean flow or coarse-grained fluid motion.

Turbulence involves random behavior, like molecular motion, so the application of similar statistical mechanics procedures seems appropriate. Navier (1823) first derived the usual Laplacian expression for viscosity by statistical modeling of "atomic" motion. At that time the concept of atoms was not established, and not clearly distinguished from dust particles or small fluid parcels.

In the twentieth century, it became clear that turbulence has specific properties, different from the thermal motion of molecules. There is a gap between the smallest scale of fluid motion and the microscopic scale of thermal fluctuations. While molecules interact only at short range, fluid parcels interact by pressure forces, which are long ranged. Turbulent motion thus involves a wide range of eddy sizes, unlike thermal motion. Therefore the separation between the large scale motion, considered as macroscopic, and the turbulent fluctuations is not clear. Another major difficulty is the irreversible behavior of fluid turbulence: the energy of motion is transferred toward small scales by cascade phenomena and dissipated by viscosity at the smallest scale of eddy motion, the so-called Kolmogoroff (Kolmogorov) scale. As the usual procedures of statistical mechanics apply close to equilibrium, their application to turbulent motion remains problematic.

These difficulties are partly overcome in two-dimensional turbulence, which is a good model for large scale atmospheric or oceanic motion. Two-dimensional turbulence is also relevant for conducting fluids or plasmas in a magnetic field. Such two dimensional fluid motion can be described in terms of interacting vortices which conserve their circulation, like a charge. The long-range interaction between vortices is indeed formally analogous to electrostatic interactions between charges. An application of equilibrium statistical mechanics to a set of singular vortices was first proposed by Onsager (1949), and further developed by Joyce and Montgomery (1973). An extension to a non-singular distribution of vorticity was proposed by Robert and Sommeria (1991) and independently by Miller (1990). The motivation was to explain the observed self-organization of the turbulent motion into steady coherent vortices. The most striking example is the Great Red Spot of Jupiter, a huge vortex persisting for more than 300 years in a very turbulent surrounding.

While two-dimensional turbulence tends to self-organize into steady flows, it is often maintained in unsteady regimes by forcing. This is the case for instance for oceanic currents. Then it is desirable to use Large Eddy Simulations: an explicit numerical model for large scales, involving a statistical model of the sub-grid, unresolved scales. A MEP principle has been proposed for this purpose by Robert and Sommeria (1992). The idea is that the unresolved turbulent motion should increase the entropy at the highest rate consistent with known constraints.

This MEP is intended as a method of research rather than an intangible law of physics. In the spirit of Jaynes (1985), the idea is to make the most objective guess of the transport by the unresolved turbulent motion, taking into account the known constraints. This guess can be progressively improved by taking into account new constraints. The opposite approach would be to sort out the chain of elementary instabilities and other elementary processes. Turbulence is not pure disorder: well defined flow phenomena can be distinguished.

However elementary they may seem, all processes rapidly become quite intricate, and some interactions remain unknown or neglected. The cumulative effects of small errors may lead to erroneous predictions if the general trends are not well represented. Any approximation to the dynamical equations is only valid for a finite time, and for long time predictions, the respect of the system properties is more important than the precision of the model. These properties involve the conservation of global quantities, like energy, but also the general trends to irreversibility, corresponding to entropy increase. The MEP approach allows to progressively improve our modeling capability, always keeping into account these global properties.

This formulation of MEP for turbulence modeling will be summarized in Sect. 7.3. This formulation provides also a new way to understand diffusion process in usual thermodynamics, so it will be introduced first in this context (next section). Finally, extensions to gravitational systems will be discussed in Sect. 7.4. Gravitational interactions, which are also long ranged, indeed possess interesting similarities with vortices. Likewise, the thermodynamics of gravitational systems is fascinating as entropy increase is an apparent source of order. This is of broad relevance for the organization of the universe and consequently for the evolution of life. Note finally that a connection has been recently pointed out between gravitational systems and the chemical interactions which tend to organize bacterial populations in clusters, the chemotactic aggregation (Chavanis 2003).

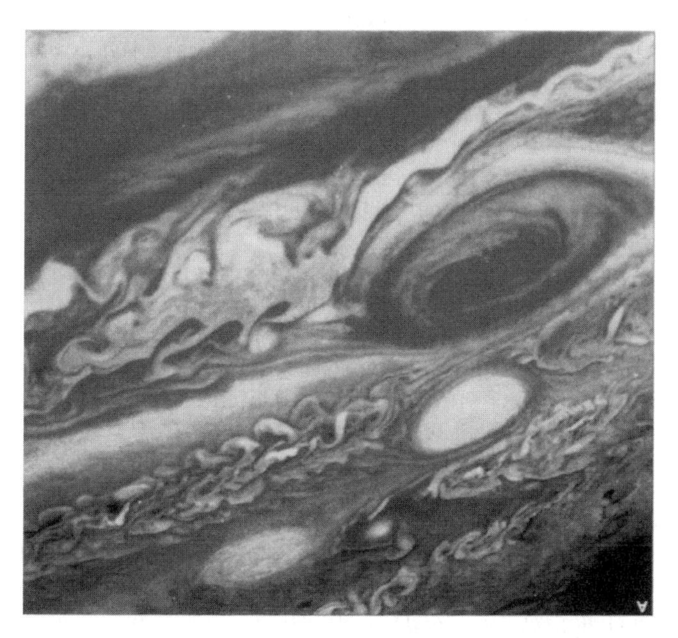

Fig. 7.1. The Great Red Spot of Jupiter (*top*) and White Oval (*bottom*) are large atmospheric vortices remaining coherent amidst turbulence. Photo: NASA

7.2 MEP in Classical Thermodynamics

The definition of entropy is subtle. It begins with a definition of elementary events, or a probability density f for continuous variables. In classical statistical mechanics, these elementary events are defined as the set of positions q and velocity $\dot{q}$ of each particle. More precisely, an event corresponds to the system being in a small interval $dqd\dot{q}$. The choice of q, $\dot{q}$ as coordinates is important. For instance, taking $q' = q^2/2$ instead of q, would yield a different probability f', such that $f(q)dq = f'(q')dq'$, so that $f'(q') = f(q)/q$. If the probability density f of q *is* uniform, the probability f' of q' is not uniform due to the different weights of the volume elements.

Therefore the probability density seems to be an arbitrary choice of our state of knowledge. It is, however, natural to choose q instead of q^2 as we expect all the positions to be equivalent, so that all the intervals dq should have the same statistical weight. Similarly we expect a priori that the two sides of a tossed coin have the same probability. Concerning velocity, the choice of a uniform probability for each component is not so obvious.

In this respect a key ingredient is the Liouville theorem, which states that the volume element in phase space, that is, the product of $dqd\dot{q}$, for each particle is conserved in time. This applies if the chosen coordinates q, $\dot{q}$ are a set of canonical variables of the Hamiltonian dynamics. The probability density in phase space is therefore locally conserved, like a dye concentration transported and stirred in an incompressible fluid. We then expect for an isolated system to reach a uniform probability among the available microscopic states. All the states of given energy tend to be reached with equal probability if there is no other constraint. This choice of density probability is justified by its consistence with the dynamical evolution. When applying the concept of entropy to more general systems, for instance turbulence, the absence of a Liouville's theorem is a fundamental difficulty, and consistency with the dynamics must be checked.

In problems of dynamical evolution, the system is assumed to be in a particular state, described by a probability distribution $f(q, \dot{q})$. Some a priori information is therefore available, quantified by the information entropy $\int f \ln f \; dqd\dot{q}$. The entropy measures the number of possible microscopic configurations consistent with the given probability density. Minimizing this entropy with the normalization constraint $\int f \ln f \; dqd\dot{q} = N$ (total number of particles) yields the uniform probability distribution. This is the state of equilibrium expected to be reached at the final stage, when all the available information on the initial state has been lost.

Usual thermodynamic systems tend to relax very quickly to a local equilibrium by molecular interactions, while the global equilibrium is reached after the much longer time scale of diffusion. Maximizing the physical entropy $S = -\int f \ln f \, d\dot{q}t$ (i.e., minimizing the information entropy $-S$) for a given local density $\rho(\mathbf{r}) = \int f \, d\dot{q}$ and local energy density $\rho(\mathbf{r})$=$(1/2)\int f \, \dot{q}^2$

$d\dot{q}$ yields the usual Gaussian distribution for velocity, characterizing a local thermodynamic equilibrium.

The slower evolution is described by the Navier-Stokes equation, which expresses the conservation of momentum. It describes the global uniformisation of density (and pressure) through acoustic wave propagation. To present MEP ideas in the absence of such complications, we shall consider that a state of uniform pressure and temperature has been reached, but a chemical has a non-uniform concentration $c(\mathbf{r})$. Then the further evolution of the system by diffusion will be associated with an increase of the sole compositional entropy

$$S = -\int c(\mathbf{r}) \ln(c(\mathbf{r})) d^3 r\,. \tag{7.1}$$

The conservation of the total mass $\int c(r)\ \mathrm{d}^3\mathrm{r}$ of the chemical is equivalent to the local expression:

$$\frac{\partial c}{\partial t} = -\nabla \cdot \mathbf{J}\,, \tag{7.2}$$

where the flux vector $\mathbf{J}$ must have a zero normal component at the edge of the system, assumed to be isolated.

We know that entropy must increase, leading to uniformisation of c. By combining (7.1) and (7.2), the rate of entropy change is

$$\dot{S} = -\int J \cdot \nabla(\ln c) d^3 r\,. \tag{7.3}$$

In the traditional thermodynamic approach for non-equilibrium, a linear relationship between the flux $\mathbf{J}$ and the concentration gradient is assumed

$$\mathbf{J} = -\kappa \nabla c\,. \tag{7.4}$$

This is usually justified as the first term in an expansion in terms of ∇c, which expresses the distance to equilibrium. The diffusion coefficient κ must be positive to assure the entropy increase. Indeed (7.3) leads to $\dot{S} = \int \kappa (\nabla c)^2\, d^3 r$.

The same result can be obtained by a MEP principle, expressed as follows for each macroscopic state of the system, characterized by $c(\mathbf{r})$, the flux $\mathbf{J}$ maximizes the entropy production (7.3) with the constraint $|\mathbf{J}| < A\ (\mathbf{r})$. We do not specify the bound A($\mathbf{r}$), it must only exist.

It can be shown that the bound on $|\mathbf{J}|$ is reached: it is more advantageous in terms of entropy production to increase $|\mathbf{J}|$, as it is obvious from the expression (7.3) of entropy production. It is also obvious that for a given modulus, $\mathbf{J}$ optimizes $\dot{S}$ when it is aligned with ∇c, so that the classical expression (7.4) is recovered.

We can derive this relationship in a more formal way, by introducing a Lagrange multiplier $\lambda\ (\mathbf{r})$ associated with the constraint on $|\mathbf{J}\ |$ at each point. Then the first variations must satisfy the relationship

$$\delta \dot{S} - \int \lambda(\mathbf{r}) \delta \mathbf{J}^2 d^3 r = 0 \,. \tag{7.5}$$

Using (7.3) this relationship writes $\int$ [∇(ln c)+ $2\lambda\mathbf{J}$] $\delta\mathbf{J} = 0$, which must be satisfied for any variation δJ. This is only possible if the integrand vanishes, leading again to (7.4) (with a diffusion coefficient κ=(2 λc)$^{-1}$).

This seems a quite natural derivation. However we must keep in mind that some assumptions are hidden. First the entropy definition relies on the choice of the particle positions q as the elementary variable. Choosing q^2 for example would lead to a uniform concentration per unit of r^2. This is justified by Liouville's theorem for a Hamiltonian system, but generalization is not obvious. Secondly we need to distinguish the fast process, and the slow process of diffusion which controls the global relaxation to equilibrium. We also neglect memory effects, assuming that the flux $\mathbf{J}$ at any time depends only on the present macroscopic state.

7.3 MEP in Two-Dimensional Turbulence

As stated in the introduction, our MEP approach was introduced in the context of 2D turbulence. This is a complex flow confined to a surface, for instance a planetary atmosphere at large scale, or a flow in a volume in the presence of constraining external force, resulting from the Coriolis or electromagnetic effects (see e.g., Sommeria (2001) for a recent review). We consider the fluid as inviscid, so the relevant equation is Euler's equation in two dimensions, and not Navier-Stokes. This is quite justified, as the solutions have been proved to remain regular for any time in two dimensions. Euler's equation is best written in terms of the vorticity field ω=(curl $\mathbf{u}$)$_z$, where z is the normal to the plane of motion:

$$\begin{aligned} \frac{\partial \omega}{\partial t} + \mathbf{u} \cdot \nabla \omega &= 0 \\ \nabla \cdot \mathbf{u} &= 0 \,. \end{aligned} \tag{7.6}$$

It expresses the material conservation of the vorticity ω by the divergenceless flow $\mathbf{u}$. The condition $\nabla \cdot \mathbf{u} = 0$ is equivalent to the statement that $\mathbf{u}$ is derived from a stream function ψ by $u = e_z \times \nabla \psi$ where e_z is the vertical unit vector. Each fluid parcel preserves its local rotation like a small gyroscope, but it is transported and stirred by the velocity field induced by all the other vorticity parcels.

The total energy of the flow is conserved. It can be expressed in terms of the vorticity ω and the stream function ψ, by using integration by parts:

$$E = \frac{1}{2} \int u^2 d^2 r = \frac{1}{2} \int \psi \omega d^2 r \,. \tag{7.7}$$

Note that the expression of energy is formally analogous to the potential energy for charge interactions with a potential ψ. This potential similarly satisfies the Poisson equation $-\Delta\psi = \omega$.

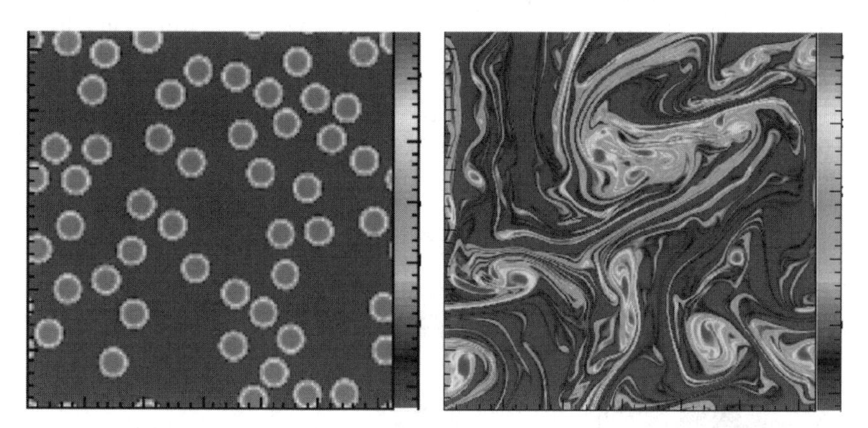

Fig. 7.2. Numerical simulation of two-dimensional turbulence, represented by vorticity maps (from Bouchet 2003). The initial state (*left*) is made of uniform vorticity patches. These patches deform into complex filaments (*right*) which tend to wrap up into large coherent vortices

It is known that vorticity is stirred at small scale, as shown in numerical simulations (e.g., Fig. 7.2). We are not interested in these fine scales but rather in the velocity field, which is much smoother, as it depends on some locally averaged vorticity. The idea developed by Robert and Sommeria (1991) and Miller (1990) is to describe the system in a macroscopic way, as the local probability density $\rho(\sigma,\mathbf{r})$ of finding the vorticity level σ in a neighborhood of the position $\mathbf{r}$. We are mostly interested in the local vorticity average $\overline{\omega}$, and the associated "coarse-grained" velocity field defined by curl $\mathbf{u} = \overline{\omega}$. As each vorticity parcel conserves its vorticity, we have the global conservation of $\gamma(\sigma) = \int \rho(\sigma, \mathbf{r})\, d^2r$ for each vorticity value σ. This conservation law can be written in a local form, by introducing a diffusion flux $\mathbf{J}(\sigma, \mathbf{r})$.

$$\frac{\partial \rho}{\partial t} + \mathbf{u} \cdot \nabla \rho = -\nabla \cdot \mathbf{J}\,. \tag{7.8}$$

Each vorticity level behaves like a chemical species: it is globally conserved but it is transported and mixed. In comparison with the diffusion (7.2), we have sorted out the transport $\mathbf{u}\cdot\nabla\rho$ by the explicitly resolved (coarse-grained) velocity $\mathbf{u}$. We limit the MEP application to the transport by the local, unknown fluctuations.

The expression of entropy is now obtained by replacing the concentration c by ρ in (7.1), and integrating over the vorticity levels σ. A weak form of a Liouville's theorem can be invoked to justify this entropy (Robert 2000). We have now two additional constraints: first the fluid parcels locally exclude each other, resulting in the local normalization $\int \rho(\sigma, \mathbf{r}) d\sigma = 1$. The other constraint is brought by energy conservation. The energy (7.7) can be written in terms of the densities by replacing ω by its local average $\overline{\omega}$.

Introducing these additional constraints, MEP yields the diffusion current

$$\mathbf{J} = -\kappa \left[\nabla \rho + \beta \rho (\sigma - \overline{\omega}) \nabla \psi \right] . \tag{7.9}$$

It contains a diffusion flux in $\nabla \rho$ and a second term in which $\nabla \psi$ acts at large scale. It is formally similar to a drift term, like for a charged particle submitted to an electric field $-\nabla \psi$. The coefficient β is determined by the condition of global energy conservation.

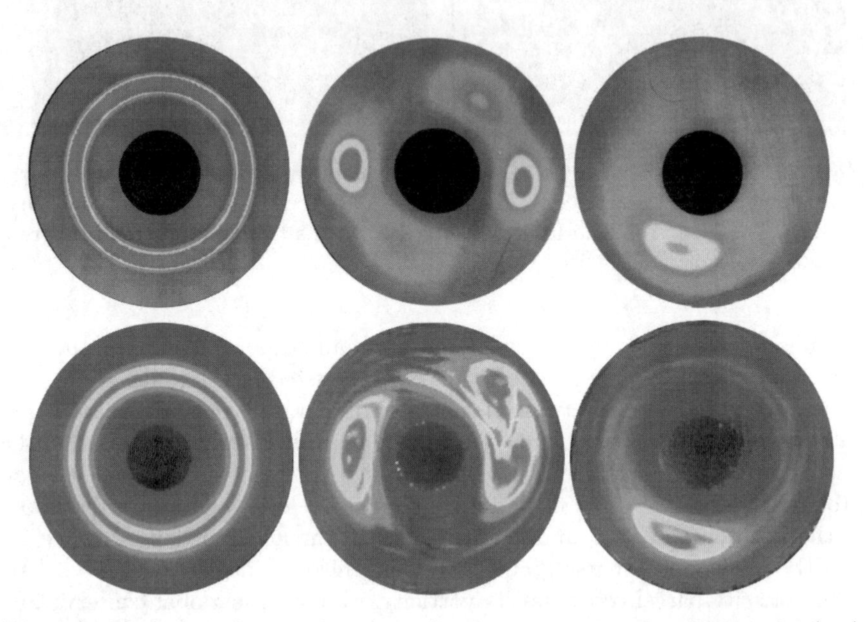

Fig. 7.3. Evolution of a two-dimensional vortex ring using the MEP model (*top*) and a high resolution numerical simulation (*bottom*). The initial ring (*left*) develops sheat instabilities (*middle*), and eventually self-organizes into a unique coherent vortex (*right*). Voriticity filaments are visible in the high resolution simulations, but the MEP diffusive flux smooth them out while preserving the organization into the final steady state. From Sommeria (2001)

At equilibrium, the two terms balance each other, and we get a steady-state solution characterized by a given relationship between the density ρ and the stream function. This represents some mean flow, for instance a large vortex, which is predicted to emerge from turbulence. It can be directly obtained as a statistical equilibrium by maximizing the entropy with given energy. This explanation of self-organization by entropy maximization may seem paradoxical, but the organization is in reality due to the constraint on energy: Full mixing of vorticity would be inconsistent with energy conservation, and the optimum state appears to be a large coherent vortex surrounded by well mixed vorticity.

In this model, we have replaced the initial vorticity equation by a set of equations for each vorticity level σ. It is still computationally advantageous as only a coarse spatial resolution is needed, thanks to the smoothing effect of the diffusive flux $\mathbf{J}$. By contrast the resolution of the initial Euler's equation requires a very high resolution to resolve the fine scale filaments. It turns out that taking into account a few vorticity levels σ is generally sufficient. Furthermore, if the initial condition is made of a patch with a single non-zero vorticity level σ_0, then the probability density $\rho(\sigma, \mathbf{r})$ is a Dirac distribution in σ, and we can identify ρ with the coarse-grained vorticity $\overline{\omega}$.

An application of this model to the organization into a single coherent vortex, like in the case of the Great Red Spot of Jupiter, is shown in Fig. 7.3. The initially developed vortices merge into a single one, which remains indefinitely as an equilibrium state. Usual turbulence models lead instead to an eventual decay of the vortex by diffusion. A specific application to the Great Red Spot is discussed by Bouchet and Sommeria (2002).

Traditional turbulence modeling relies on closure approximations: hypothesis on the probability laws are assumed, consisting in neglecting some correlations. Kinetic models of molecular motion, like the Boltzmann or the Fokker-Planck equations, also rely on a similar closure approximation. A derivation of a kinetic transport equation for two-dimensional turbulence has been proposed by Chavanis (2000). This approach confirms the general form of the MEP result given by equation (7.9): transport by velocity fluctuations involves a usual diffusion term, plus a long range "drift term". Its physical origin is interpreted as a long range "polarization" of vorticity by the influence of local vorticity fluctuations. The expression of this drift term is, however, more complex, and memory effects cannot be neglected: the polarization results from the history of flow straining.

Closure models and the MEP approach are complementary. MEP provides models consistent with the long time trends of the dynamics, but relying on some guessed constraints. Furthermore the values of the diffusion coefficients are not given by such thermodynamic approach. By contrast, closure yields these coefficients, and relies on systematic approximations, which can be at least justified for short time scales. Ideally, a good model should be a closure fully consistent with the MEP. This has not been really achieved for two-dimensional turbulence. For instance the above mentioned closure model of Chavanis (2000) does not conserve energy on long time scales. Some improvements have been also proposed for MEP, introducing the constraint of energy conservation in a local way (Chavanis and Sommeria 1997). Another improvement, by Bouchet (2003), is to consider that part of the fluctuations is not random but results from the systematic straining by the coarse-grained motion. Then MEP is used only for a remaining random component. There is probably not a unique answer to turbulence modeling, but rather a set of models suitable to respect some properties of the system, with a tradeoff between accuracy and complexity.

7.4 Application to Stellar Systems

We have already discussed above analogies between vortices and electrostatic charge interactions. Gravitational systems are similarly controlled by long-range interactions. We distinguish two kinds of gravitational systems, whether they are collisional or not. The first case is an ordinary self-gravitating gas, for instance during star formation. The second case corresponds to stellar dynamics in a galaxy. Then the individual "molecules" are stars.

The general trend of such systems is to form a dense core by gravitational collapse, while the released energy heats the envelope. This collapse globally increases the entropy of the system. Thus the whole process of stellar evolution can be qualitatively understood in terms of entropy increase: the initial star formation from a dilute cloud, its further evolution into a denser and denser body, associated with the expulsion of a hot gas, the solar wind or the explosion of a supernova for massive stars. These processes are well described by the fluid dynamics of compressible gas, and MEP does not seem to be of much use in this stellar case.

Fig. 7.4. Elliptical galaxies (here NGC3379) contain stars in a state of statistical equilibrium, which results from a global, fluid like, mixing in phase space. From http://www.licha.de/AstroWeb

For galaxies, the dynamics are quite different, because of the very low probability of binary interactions between stars ("collisions"). Nevertheless a strong tendency to reach statistical equilibrium is observed. In particular, for elliptical galaxies (see Fig. 7.4) the probability distribution of velocity and radial density profile fit very well the prediction of statistical equilibrium (except at the periphery). According to estimates on binary star interactions such equilibrium would be reached in most cases on a time scale greater than the age of the universe.

An explanation of this paradox has been given by Lynden-Bell (1967) by introducing the notion of violent relaxation, a kind of turbulent behavior for a fluid in the six dimensional phase space of position and velocity. This fluid satisfies the Vlasov equation which has some similarities with the Euler equation for two-dimensional turbulence. Lynden-Bell (1967) proposed a similar statistical approach as described in previous section for two-dimensional turbulence.

The application of MEP to this problem was proposed by Chavanis et al. (1996); see also Chavanis (2002, 2003). The turbulent mixing for density is similarly described by a diffusion term and a drift term proportional to the gravity field. The degree of validity of this model for actual gravitational systems is still unclear.

Remarkably similar equations arise for the chemotactic aggregation of bacterial populations (Chavanis 2003). In that case the potential corresponds to the concentration of chemicals emitted by the bacteria. The balance between emission and diffusion yields a Poisson equation, like for the potential of gravity.

7.5 Conclusions

Diffusion equations, widely used in turbulent modeling, can be derived from a MEP principle. The diffusion of a quantity can be viewed as the process which maximizes the entropy production with the natural constraint of a bounded flux for this quantity. Note that this principle should not be confused with the principle of minimum entropy production proposed by Prigogine in 1947 (see Prigogine, 1967). The later applies to the solution of known transport equations for a system submitted to external fluxes. The MEP principle is quite different, and it is designed to guess the equations of transport. The MEP principle discussed here is also different from what has been proposed for describing the heat transport in planetary atmospheres (e.g. Paltridge, this volume; Ito and Kleidon, this volume; Lorenz, this volume). In particular we are considering a dynamical entropy describing turbulent fluctuations rather than the usual thermodynamical entropy. It is not clear whether a more unified principle can arise. Furthermore, MEP, as discussed here, cannot be viewed as a law of physics in the usual sense, with a formula that we could apply and check. It is rather a guideline for modeling complex systems, possibly learning from trial and error, in the spirit of Jaynes' ideas. It does not replace a more detailed analysis of the system, but it provides an objective way of using available information.

We can refine the model by adding new constraints as we better understand the behavior of the system. A first constraint is provided by conservation laws, in particular for energy. In the case of long-range interactions, either for vortices or gravitational systems, this results in an additional drift term, which leads to self-organisation into large structures, coherent vortices or galaxies. At a next step, we can distinguish some deterministic transport effects and restrict the statistical description to a smaller random contribution. This refinement respects the general trend of entropy increase, and the known constraints of the system. This is unlike the usual approximation procedures, turbulent closure or kinetic models, which, although improving short term predictions, are prone to progressively drift away from reality on long time scales.

References

Bouchet F, Sommeria J (2002) Emergence of intense jets and Jupiter Great Red Spot as maximum entropy structures. J Fluid Mech 464: 165–207.

Bouchet F (2003) Parametrisation of two-dimensional turbulence using an anisotropic maximum entropy production principle. Phys Fluids, submitted.

Chavanis PH, Sommeria J, Robert R (1996) Statistical Mechanics of Two-dimensional Vortices and Collisionless Stellar Systems. Astrophys J 471: 385.

Chavanis PH, Sommeria J (1997) Thermodynamical approach for small scale parametrisation in 2D turbulence. Phys Rev Lett 78: 3302–3305.

Chavanis PH (2000) Quasi-linear theory of the 2D Euler equation. Phys Rev Lett 84: 5512–5515.

Chavanis PH (2002) Statistical mechanics of two-dimensional vortices and stellar systems, in Dauxois T (eds) Dynamics and thermodynamics of systems with long range interactions, Lecture Notes in Physics, Springer Verlag, Berlin.

Chavanis PH (2003) Generalized thermodynamics and Fokker-Planck equations: applications to stellar dynamics and two-dimensional turbulence. Phys Rev E 68: 036108.

Jaynes ET (1985) Where do we go from here? in: Ray Smith C, Grandy WT (eds) Maximum entropy and Bayesian methods in inverse problems, Reidel, Dordrecht, Holland.

Joyce G, Montgomery D (1973) Negative temperature states for the two-dimensional guiding center plasma. J Plasma Physics 10: 107–121.

Lynden-Bell D (1967) Statistical mechanics of violent relaxation in stellar systems. Month Notes Roy Astron Soc 136: 101–121.

Miller J (1990) Statistical mechanics of Euler equations in two dimensions. Phys Rev Lett 65: 2137–2140.

Navier CLMH (1823) Mémoire sur les lois du mouvement des fluides. Mém Acad Roy Sci 6: 389–440.

Onsager L (1949) Statistical Hydrodynamics. Nuovo Cimento Suppl 6: 279–287.

Prigogine I (1967) Introduction to thermodynamics of irreversible processes. Wiley, New York.
Robert R, Sommeria J (1991) Statistical equilibrium states for two dimensional flows. J Fluid Mech 229: 291–310.
Robert R, Sommeria J (1992) Relaxation towards a statistical equilibrium state in two-dimensional perfect fluid dynamics. Phys Rev Lett 69: 2776–2779.
Robert R (2000) On the statistical mechanics of 2D Euler and Vlasov Poison equations. Comm Math Phys 212: 245–256.
Sommeria J (2001) Two-dimensional turbulence. in: Lesieur M, Yaglom A, David F, New trends in turbulence, EDP/Springer, 387–447.

8 Entropy Production of Atmospheric Heat Transport

Takamitsu Ito[1] and Axel Kleidon[2]

[1] Program in Atmospheres, Oceans and Climate, Massachusetts Institute of Technology, Cambridge, MA 02139, USA
[2] Department of Geography and Earth System Science Interdisciplinary Center, 2181 Lefrak Hall, University of Maryland, College Park, MD 20742, USA

Summary. We examine the rate of entropy produced by the atmospheric general circulation and the hypothesis that it adjusts itself towards a macroscopic state of maximum entropy production. First, we briefly review thermodynamics of a zonally-averaged, dry atmosphere. We examine the entropy balance of a dry atmosphere, and identify the key processes that lead to entropy production. Frictional dissipation and diabatic eddy transfer are the major sources of entropy production, and both processes are dominated by baroclinic eddies in the middle latitudes. Secondly, we derive a simple solution for the upper bound of entropy production from the energy balance constraint, which can be compared to the simulated temperature distribution simulated by an idealized GCM. These temperatures agree well with the MEP solution in the mid-latitude troposphere. However, there are significant differences in tropics where the Hadley circulation controls the large-scale temperature distribution. Finally, we show that the simulated entropy production is sensitive to model resolution and the intensity of boundary layer friction, and explore the significance of dynamical constraints. We close with a discussion of the implications of the MEP state for global climatology.

8.1 Introduction

The atmospheric circulation is driven by the temperature gradient $\Delta T_{E,P}$ between the equatorial and polar regions as a result of differences in solar irradiation. This temperature gradient is not fixed, but is affected by the amount of heat transport associated with the atmospheric circulation. The transport of heat from the warmer tropics to the colder poles leads to entropy production. Paltridge (1975) first suggested that the atmospheric circulation adjusts itself to a macroscopic state of maximum entropy production (MEP). Several authors applied the MEP hypothesis to energy balance climate models (e.g., Paltridge 1978); Nicolis et al. 1980; Grassl 1981; Pujol and Llebot 2000). They suggest that the MEP solutions are in plausible agreement with observed variables characterizing the zonal mean present-day climate. Lorenz (2001) suggests that the MEP also applies to the atmospheres of other planets such as Mars and Titan (see also Lorenz, this volume). While empirical support for the MEP hypothesis has been accumulating, the fundamental mechanisms are not yet fully understood.

Lorenz (1960) suggested that the atmosphere maximizes the production rate of available potential energy (APE), which is essentially equivalent to the MEP hypothesis of Paltridge when appropriate definition of the entropy production is considered (Ozawa et al. 2003). Lorenz's hypothesis of the maximum APE production considers the general circulation of the atmosphere as a heat engine of maximum efficiency in which the production of mechanical work is maximized for given solar forcing. In a statistical steady state, the production rate of APE must balance the rate of dissipation. Ozawa et al. (2003) derive a simple linear relationship between the production rate of APE and the entropy production due to the turbulent dissipation.

Recently, Dewar (2003) studied the theoretical basis for the MEP hypothesis based on the statistical mechanics of open, non-equilibrium systems. The state of MEP emerges as the statistical behavior of the macroscopic state when the information entropy is maximized subject to the imposed constraints. Dewar's theory is generally applicable to a broad class of the steady state, non-equilibrium system, such as fluid turbulence. Studies of Paltridge and others could be considered as a particular representation of the MEP principle in the climate system.

The MEP hypothesis has been applied to different types of climate models with various assumptions. Shutts (1981) applied the MEP hypothesis to the two-layer quasi-geostrophic model, and maximized the entropy production with the constraints of energy and enstrophy conservation. The extremal solution of Shutts is somewhat comparable to the ocean gyre circulation. Ozawa and Ohmura (1997) applied the MEP hypothesis to radiative convective equilibrium model, and reproduced a reasonable vertical temperature profile associated with the MEP state. Shimokawa and Ozawa (2002, also this volume) examined the entropy production in the multiple steady states of an ocean general circulation model, and suggest that the system tends to be more stable at higher rates of entropy production. Kleidon (2004) showed with a simple two box model of the surface-atmosphere system that the partitioning of energy at the Earth's surface into radiative and turbulent fluxes can also be understood by MEP. The MEP hypothesis was also recently demonstrated to emerge from atmospheric General Circulation Model (GCM) simulations in which model resolution and boundary layer friction was modified (Kleidon et al. 2003). These results may also serve as empirical supports for the basic concept of the MEP in the climate system.

This chapter investigates the physical processes that control entropy production in the atmospheric general circulation and how they may be related to the MEP hypothesis. In particular, we use atmospheric GCMs to simulate large-scale eddies in the atmosphere and examine their role in setting the atmospheric heat transport and the associated entropy production. We first review the entropy balance in the zonally-averaged dry atmosphere and consider the entropy balance in the model. Secondly, we analytically derive the upper bound of entropy production, and examine how close the simulated entropy production is to the theoretical upper bound. We also compare the temperature distribution of the analytic MEP solution to that of the numerical simulation and discuss the importance of dynamical constraints, imposed by angular momentum conservation. We then show that simulated entropy production is sensitive to model resolution and the intensity of boundary layer friction, and shows a characteristic maximum. We close with a discussion of the implications of the MEP state for climatology.

8.2 Entropy Production in an Idealized Dry Atmosphere

We briefly review the thermodynamic balance of a zonally-averaged dry atmosphere. A change in specific entropy, ds, of an air parcel with temperature T is defined as $Tds = dQ$. A detailed derivation of atmospheric thermodynamics can for example be found in Gill (1982). Following the trajectory of an air parcel, the change in the specific entropy is related to the rate of diabatic heating, DQ/DT:

$$\frac{Ds}{Dt} = \frac{1}{T}\frac{DQ}{Dt} \tag{8.1}$$

The potential temperature θ of the air parcel can be defined in terms of its specific entropy, $s = c_p \ln \theta$, where c_p is specific heat of dry air at constant pressure. The thermodynamic equation can then be derived from 8.1:

$$\frac{D\theta}{Dt} = \frac{1}{c_p}\frac{\theta}{T}\frac{DQ}{Dt} \tag{8.2}$$

Adiabatic processes conserve both specific entropy and potential temperature. For a dry atmosphere, the diabatic heating term includes heating due to thermal diffusion, viscous dissipation, and radiative fluxes. We parameterize the radiative heating and the frictional dissipation as in Held and Suarez (1994) (hereafter, HS94), which is often used to evaluate the hydrodynamics of atmospheric general circulation and climate models. The radiative transfer is parameterized as a Newtonian damping of local temperature to the prescribed radiative-convective equilibrium profile T_{eq}:

$$\frac{DQ}{DT} = -c_p k_T (T - T_{eq}) + k_U (u^2 + v^2) \tag{8.3}$$

The second term on r.h.s. represents the heating due to viscous dissipation, which is often neglected since its magnitude is very small (a few percent) compared to that of the radiative heating. Here, we include this term for consistency in the energy balance. The frictional damping coefficient k_U, the radiative cooling coefficient k_T, and the radiative-convective equilibrium temperature profile T_{eq}, are prescribed functions of latitude and pressure. For the detailed distribution of T_{eq}, k_T and k_U, see HS94.

Next, we zonally average the thermodynamic equation (8.2). It is convenient to define the *Exner* function $\pi = (p/p_0)^\kappa$, where p_0 is the surface pressure and $\kappa = 2/7$. With this definition, we can express the thermodynamic equation as

$$\frac{\partial \overline{\theta}}{\partial t} + \nabla \cdot \overline{u}\,\overline{\theta} = -\nabla \cdot \overline{u'\theta'} + \frac{1}{c_p \pi(p)} \overline{\frac{DQ}{Dt}} \tag{8.4}$$

Here, zonally averaged quantities are overlined such as $\overline{A}$, and the deviations from the mean are written as $A' = A - \overline{A}$. We obtain the zonal mean entropy balance equation by multiplying both sides of 8.4 by $c_p/\overline{\theta}$:

$$\frac{\partial \overline{s}}{\partial t} + \nabla \cdot \overline{u}\,\overline{s} = -c_p \nabla \cdot \left(\frac{\overline{u'\theta'}}{\overline{\theta}} \right) - c_p \frac{\overline{u'\theta'} \cdot \nabla \overline{\theta}}{\overline{\theta}^2} + \frac{1}{T} \overline{\frac{DQ}{Dt}} \tag{8.5}$$

The r.h.s. of this equation contains two components of eddy fluxes that contribute to entropy production. The first term on the r.h.s. represents the adiabatic component of the eddy transfer which vanishes when integrated globally. The second term is the diabatic component of the eddy transfer which does not vanish when integrated globally.

8.2.1 Global Budget of Energy and Entropy

Globally integrated, the radiative heating must be zero for a steady state. Thus, we have

$$< c_p\, k_T (T - T_{eq}) >= 0 \tag{8.6}$$

with the brackets denoting the global integral, $<>= -\int dx\ dy\ dp/g$. Combining (8.3) and (8.5) and integrating globally, we derive the global entropy balance:

$$< \sigma_{TOT} >= \left\langle \frac{c_p k_T (\overline{T} - T_{eq})}{\overline{T}} \right\rangle = \left\langle -c_p \frac{\overline{u'\theta'} \cdot \nabla \overline{\theta}}{\overline{\theta}^2} + \frac{k_U (\overline{u^2 + v^2})}{T} \right\rangle \tag{8.7}$$

The globally integrated entropy production $<\sigma_{TOT}>$ can be expressed in terms of the net outgoing entropy flux, and it is balanced by the integral of local entropy production through diabatic eddy fluxes and frictional dissipation. This particular formulation does not involve entropy production due to radiative transfer, dry and moist convection and other moist processes (see e.g., Pauluis, this volume). For example, (8.7) implies that $<\sigma_{TOT}> = 0$ when the atmosphere is in radiative-convective equilibrium $T = T_{eq}$. This definition of entropy production is essentially identical to the definition of Paltridge (1975).

8.2.2 Sources of Entropy Production

The atmospheric general circulation has internal sources of entropy due to dissipative processes. We diagnose the simulated fields, quantify the spatial distribution of these entropy sources, and illustrate the dynamical process controlling the entropy production. Frictional dissipation and diabatic eddy fluxes can be diagnosed directly from the simulated fields. We can define a local entropy production, which includes a frictional component σ_{fric} and an eddy component σ_{eddy}:

$$\sigma_{fric} = \frac{k_U(\overline{u}^2 + \overline{v}^2 + \overline{u'^2} + \overline{v'^2})}{\overline{T}} \tag{8.8}$$

$$\sigma_{eddy} = -c_p \frac{\overline{u'\theta'} \cdot \nabla\overline{\theta}}{\overline{\theta}^2} \tag{8.9}$$

The units of σ_{fric} and σ_{eddy} are $\mathrm{W\,K^{-1}\,kg^{-1}}$, and they satisfy $\sigma_{TOT} = \sigma_{fric} + \sigma_{eddy}$ (see 8.7). In the following, we discuss entropy production rates per unit area, that is, integrated over the vertical column of air and in units of $\mathrm{mW\,m^{-2}\,K^{-1}}$, and not per unit weight.

8.2.3 Theoretical Upper Bound of Entropy Production

In order to derive an upper bound of the global rate of entropy production σ_{TOT}, as described by (8.7), we maximize σ_{TOT} subject to the constraint of global energy balance, as described by (8.6). Dynamical constraints, as for instance imposed by the conservation of angular momentum, are not included in the constraint, so the model can be considered to be an energy balance model. We introduce a Lagrange multiplier μ and combine the constraint (8.6) and the cost function $< \sigma_{TOT} >$:

$$< \sigma_{TOT} > = \left\langle \frac{c_p k_T(\overline{T} - T_{eq})}{\overline{T}} \right\rangle + \mu \langle c_p k_T(\overline{T} - T_{eq}) \rangle \tag{8.10}$$

The rate of entropy production $< \sigma_{TOT} >$ is then extremized by setting $\delta < \sigma_{TOT} > = 0$. The resulting Euler-Lagrange equation for this extremization is:

$$T_{MEP} = \mu^{-1/2} {T_{eq}}^{1/2} \tag{8.11}$$

T_{MEP} is the temperature distribution associated with the upper bound of entropy production. The Lagrange multiplier μ is calculated by combining (8.6) and (8.11):

$$\mu^{-1/2} = < k_T T_{eq} > / < k_T {T_{eq}}^{1/2} > \tag{8.12}$$

The maximum in entropy production is then calculated by using (8.10):

$$< \sigma_{\text{TOT,MEP}} > = < c_p\, k_T(1 - \mu^{1/2} {T_{eq}}^{1/2}) > \tag{8.13}$$

Applying definitions of k_T and T_{eq} following HS94, we find $\sigma_{\text{TOT}} \approx 8.4\,\mathrm{mW\,m^{-2}\,K^{-1}}$. The resulting temperature profile T_{MEP} is shown in Fig. 8.1.

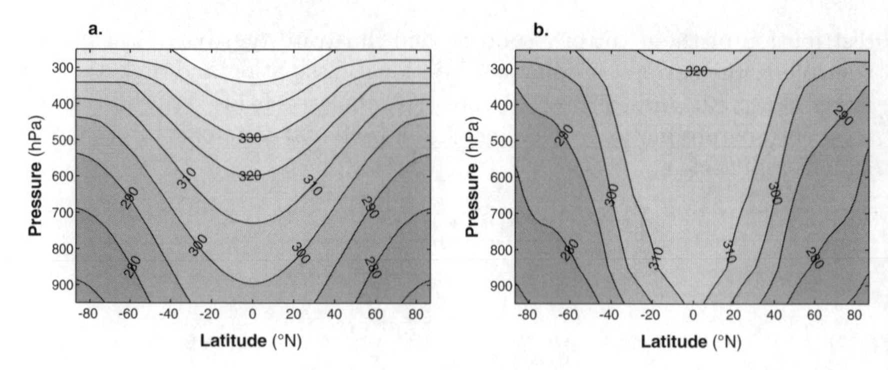

Fig. 8.1. Distribution of zonally-averaged potential temperature **a** resulting from a state of maximum entropy production derived analytically and **b** simulated with a GCM

8.3 Testing Maximum Entropy Production with Atmospheric General Circulation Models

The analytic form of the MEP solution given by (8.13) and the associated temperature distribution given by (8.11) is compared with the simulated properties of an atmospheric GCM. We first present the simulated entropy production of an atmospheric general circulation model including its latitudinal variation. Next, we compare the simulated temperatures to those associated with the theoretical upper bound of entropy production. In the third part we then discuss the sensitivity of simulated entropy production to model resolution and boundary layer turbulence in order to illustrate the conditions for MEP states associated with the atmospheric circulation.

8.3.1 Simulated Entropy Production in the Climatological Mean

The GCM we use consists of the hydrodynamical core of MITgcm (Marshall et al. 1997a,b) with idealized thermodynamics. Diabatic heating is parameterized as a Newtonian restoring term (HS94). The model does not include radiative transfer calculations or the water balance. The hydrodynamic core is able to resolve mid-latitude baroclinic eddies which play a dominant role in heat transport in the atmosphere. We consider the statistical mean state of the simulated atmosphere. Figure 8.1b shows the temporally and zonally averaged distribution of temperature of the model.

Figure 8.2 shows the distribution of vertically integrated σ_{fric} and σ_{eddy}. First, we consider the hemispheric distribution of σ_{fric}. In each hemisphere, there is a smaller peak in the tropics and a larger in the mid-latitudes. The dissipation of the mean kinetic energy is responsible for the smaller peak in tropics. The greater peak in the mid-latitudes is due to the damping of the eddy kinetic energy, which dominates global frictional entropy production

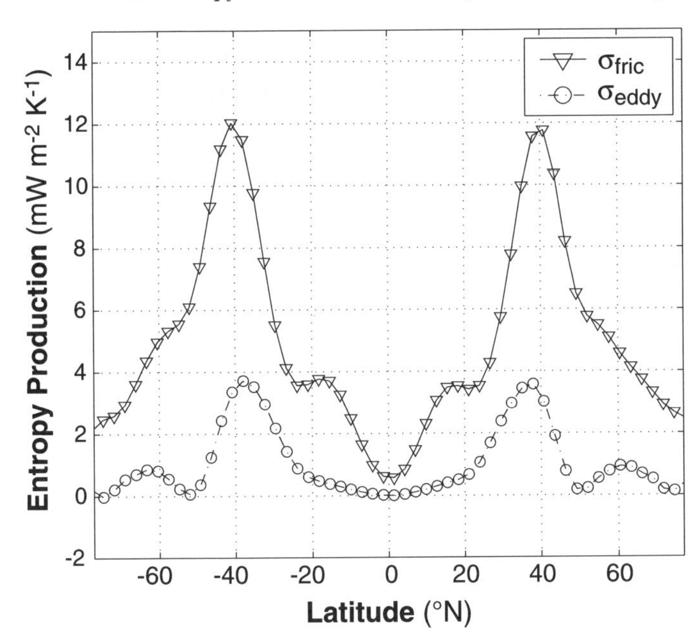

Fig. 8.2. Distribution of zonally averaged entropy production. Triangles and circles represent the frictional and eddy component of entropy production respectively

$< \sigma_{fric} >$. Observations estimate entropy production by friction to be about 6.5 mW $m^{-2}\,K^{-1}$ (Peixoto et al. 1991; Goody 2000) which compares well with the slightly smaller simulated value of 5.0 mW $m^{-2}\,K^{-1}$.

Entropy production by diabatic eddy transfer has a maximum in the mid-latitudes. The magnitude of σ_{eddy} is much smaller than σ_{fric}, suggesting that diabatic eddy fluxes plays rather minor role in the global entropy production. The magnitude of $< \sigma_{eddy} >$ is approximately 1 mW $m^{-2}\,K^{-1}$. Both σ_{eddy} and σ_{fric} have peaks around 35 N and 35 S, reflecting the significant role of baroclinic eddy transfer in controlling the magnitudes and spatial distribution of entropy production. Combined, the total entropy production in the model is about 6 mW $m^{-2}\,K^{-1}$.

8.3.2 Comparing the Analytic MEP Solution to the Simulated Atmosphere

We test the MEP hypothesis by comparing the simulated temperature distribution and meridional heat transport to those of the analytic MEP solution (Fig. 8.1 and Fig. 8.3). The analytic MEP solution has qualitative similarities to the simulated profile in the mid- and high latitudes. Surface equator-pole temperature difference is in the order of 35 K in both the MEP solution and the modeled atmosphere. Given the simplicity of the MEP solution in (8.11),

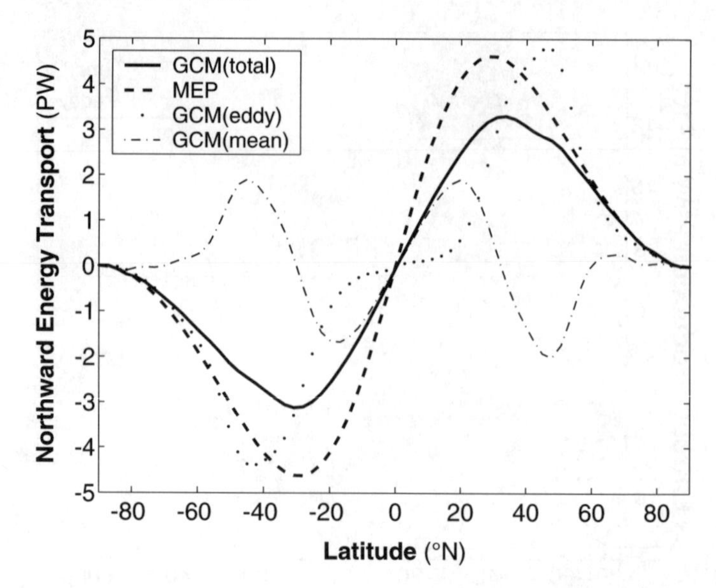

Fig. 8.3. Meridional heat transport of the analytic solution in comparison to the simulated components of the GCM

it is rather remarkable that it can capture the gross measure of the large-scale temperature gradient and heat transport.

However, there is a large disagreement in the temperature in the tropics where the simulated potential temperature field shows a uniform distribution. In the upper tropical atmosphere, the Hadley circulation dominates the temperature distribution and energy transport. The disagreement may result from the lack of dynamical constraints in the MEP solution. The MEP solution derived here is based on the energy balance constraint only, but it has been shown that the momentum balance (which is not included) is essential for the dynamics of Hadley cell (Held and Hou 1980).

In Fig. 8.4, we evaluate the MEP solution in terms of the square-root relationship between T and T_{eq}. The MEP solution in (8.11) suggests that the zonally averaged absolute temperature is proportional to the square root of the radiative equilibrium temperature profile. To test this scaling relationship, we plot T and T_{eq} in logarithmic scale. The solid line with a slope of 0.5 represents the scaling from the MEP solution. Temperature of low and high altitudes are plotted separately in order to show the qualitative differences between the tropics and high latitudes. The MEP solution compares better with the temperature of high latitudes where T is greater than 260 K. The cold temperatures of the atmosphere in low latitudes tend to have greater slope than 0.5.

The globally integrated entropy production of the simulated climate $< \sigma_{TOT} >$ is about 71% of $< \sigma_{TOT,MEP} >$. It is reasonable that the simulated entropy production is somewhat less than the upper bound because of

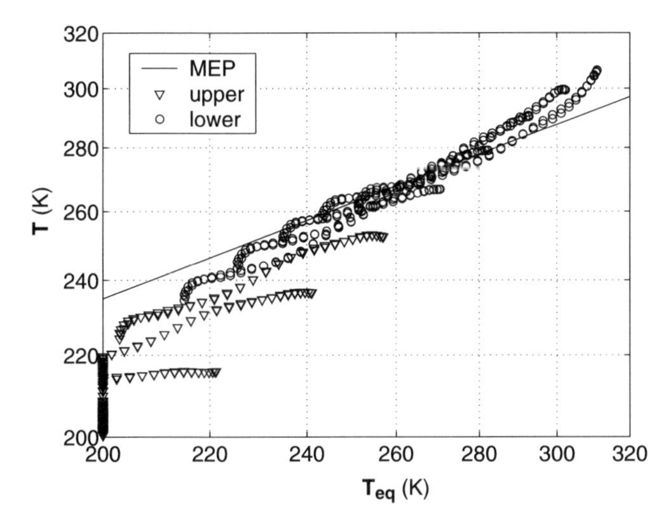

Fig. 8.4. Square root relationship between the T_{eq} and T as expected from MEP. The *solid line* represents the theoretical MEP solution with a slope of 0.5. The circles and triangles represent simulated temperatures from the lower (< 500 mb) and upper (> 500 mb) respectively

the lack of inclusion of the dynamical constraints. Frictional dissipation is the dominant source of entropy production, contributing approximately 83% to the total, with the remaining 17% originating from thermal dissipation. The integral balance of (8.10) does not exactly hold in the simulation because of spurious source of entropy from numerical diffusion. We find that the entropy production due to numerical dissipation becomes small when the horizontal and vertical resolution of the model is sufficiently high.

8.3.3 Sensitivity of Entropy Production to Internal Parameters

In order to understand why the atmospheric circulation would adjust to a state close to MEP, Kleidon et al. (2003) conducted sensitivity simulations with an atmospheric general circulation model similar to the one discussed above (Fraedrich et al. 1998, available for download at http://puma.dkrz.de). Two different methods are used in GCMs to represent turbulent processes such as the development of large-scale eddies in the mid-latitudes and the vertical circulations in the boundary layer at much finer scales.

The dynamics of mid-latitude turbulent mixing is explicitly resolved by the model. However, the spatial resolution of the model is externally prescribed and sets lower limits on the spatial structure of large-scale eddies that can be simulated. Higher model resolutions permit finer structures of the atmospheric circulation, increasing the potential number of modes (or degrees of freedom). Following Dewar (2003, also this volume), we should therefore expect an increase in entropy production with model resolution until sufficiently high degrees of freedom are allowed for by the model resolution. This increase of entropy production up to a certain level is found in the model sensitivity simulations in which the spatial resolution is varied (Fig. 8.5a).

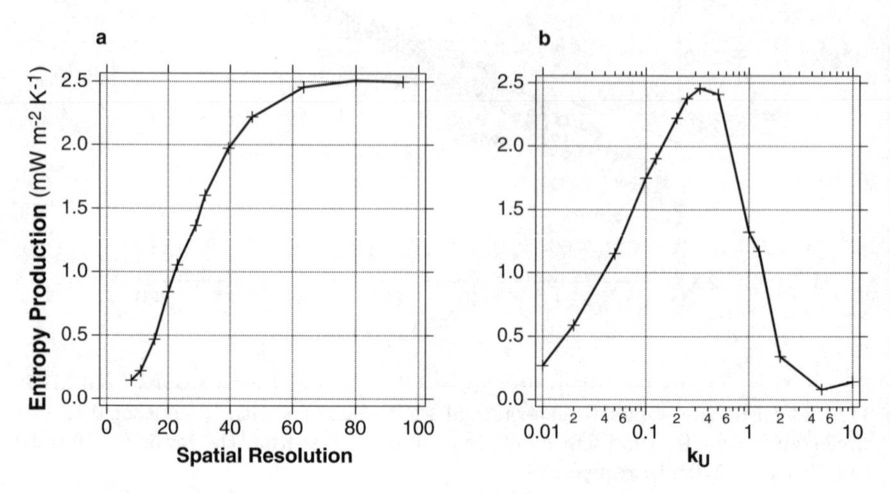

Fig. 8.5. Sensitivity of simulated total entropy production associated with atmospheric heat transport to **a** the model's spatial resolution (expressed by the number of latitudinal bands, with higher values representing finer resolution) and **b** to the frictional coefficient (with higher values representing increased boundary layer turbulence). After Kleidon et al. (2003)

The boundary layer turbulence that develops as a result of surface friction occurs at a much finer scale than GCMs are able to resolve. The effect of boundary layer turbulence on the energy and momentum balance is commonly parameterized in a fairly simple manner. In the idealized GCMs considered here, it is crudely parameterized as a Rayleigh friction term (represented by k_U in 8.3, or described by a friction time scale τ_{FRIC}). For this type of turbulence parameterization, entropy production shows a maximum (Fig. 8.5b), similar to the simple two-box energy balance example which is used to demonstrate the existence of a MEP state (Fig. 1.4). Note that the analytic form of the MEP solution in (8.11) is not sensitive to this parameter since the maximization does not explicitly include the dynamical constraints of momentum conservation.

The maximum in entropy production in Fig. 8.5 originates from the competing effects of boundary layer turbulence on eddy activity (James and Gray 1986): At the high friction extreme, momentum is rapidly removed, therefore preventing substantial eddy activity. With the reduction in friction intensity, the atmospheric flow becomes increasingly zonal, and therefore more stable to baroclinic disturbances. Consequently, the peak in entropy production corresponds to a maximum in baroclinic activity in the model. (It should be noted in the discussion above that the model does not distinguish between boundary layer turbulence and surface friction. Therefore, the sensitivity to friction should be interpreted as a sensitivity to the characteristics of boundary layer turbulence, and not surface friction per se.) Also note that the rates of entropy production in Kleidon et al. (2003) and shown in Fig. 8.5 are less than the ones obtained above which is likely due to the fact that diabatic heating by friction is not included in their model formulation.

8.4 Climatological Implications

In this chapter we have reviewed the thermodynamics of the dry atmospheric circulation, derived a temperature distribution corresponding to a state of MEP, and showed that the simulated temperature fields of an atmospheric general circulation model is broadly similar to the theoretical derived value. Naturally, the considerations used here are subject to some limitations. Most importantly, our focus on the dry atmosphere is limited with respect to the Hadley circulation, since it is driven by the latent heat flux, and therefore explicitly by moist diabatic processes. These processes contribute considerably to the overall entropy production (Pauluis and Held 2002a,b; also Pauluis, this volume). Our theoretical derivation did not include the conservation of potential vorticity, which is consistent with the fact that the simulated entropy production by the GCM is less than the theoretical estimate. Considering the conservation of potential vorticity is also likely to be important when climates of planets with different rotation rates are considered (which will affect the sensitivity of entropy production to boundary layer turbulence as discussed in the previous section). The important role of the oceanic circulation in contributing to the overall heat transport may also play a role in the distribution of temperature and the state of MEP, but has not been considered here. With these limitations in mind, we nevertheless demonstrated the important role of baroclinic activity for entropy production associated with frictional dissipation in the planetary boundary layer and mixing of warm and cold air masses in the mid-latitudes.

Furthermore, we have shown that the state of MEP as simulated by the simple GCMs used here is sensitive to the model parameterization of boundary layer turbulence and model resolution. If we take the state of MEP as representative of the macroscopic steady-state atmospheric circulation, then these sensitivities can have important implications for the application of GCMs to climate research. Since MEP represents the state of highest baroclinic activity, it also is associated with the most effective heat transport to the poles. This leads to the least temperature gradient $\Delta T_{E,P}$ for the simulation with MEP in comparison to simulations with lower model resolution or other intensities of boundary layer friction (Fig. 8.6). These model results suggest that in comparison to MEP, any other macroscopic state of the atmospheric circulation would show less baroclinic activity, and therefore

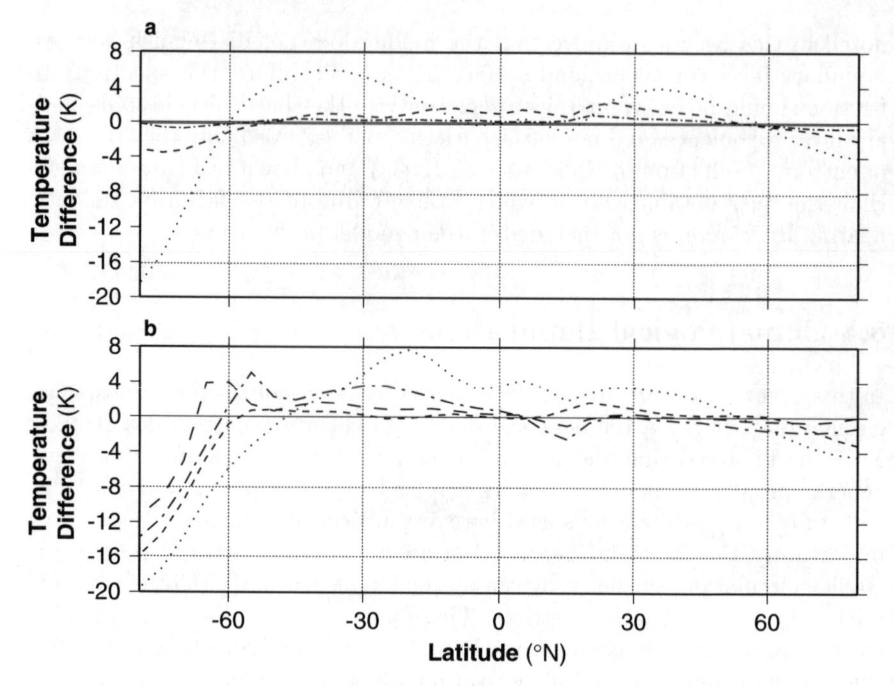

Fig. 8.6. Difference in the latitudinal variation of temperatures for the lowest atmospheric model layer in comparison to the simulated climate of maximum entropy production (for a southern hemisphere winter setup). **a** effects of different model resolutions between T10 and T42 resolution (*dotted*), same for T21 (*dashed*), and T31 (*dash-dotted*) resolution, each with optimum values of τ_{FRIC}. **b** effects of different intensities of boundary layer turbulence between $\tau_{FRIC} = 0.1$ day and $\tau_{FRIC} = 3$ days (*dotted*), same for $\tau_{FRIC} = 1$ day (*short dashes*), same for $\tau_{FRIC} = 10$ days (*long dashes*), same for $\tau_{FRIC} = 100$ days (dash-dotted), each at T42 resolution. After Kleidon et al. (2003)

transport less heat to the poles, leading to an overestimation of the equator-pole temperature gradient. This in turn may have important consequences for the adequate simulation of climatic change. It is generally known that GCMs tend to overestimate $\Delta T_{E,P}$ in paleoclimatology, for instance during periods of high carbon dioxide concentrations of the Eocene (Pierrehumbert 2002). Following the line of reasoning presented here, this may simply be an artifact of a GCM setup which does not represent a MEP climate.

Acknowledgements. T. I. would like to thank John Marshall for introducing the author to the MEP hypothesis, fruitful discussion and helpful comments. A. K. thanks Klaus Fraedrich and his research group for stimulating discussions and hospitality during the author's stay in July 2003. A. K. gratefully acknowledges financial support from the National Science Foundation though grant ATM-0336555.

References

Dewar RC (2003) Information theory explanation of the fluctuation theorem, maximum entropy production, and self-organized criticality in non-equilibrium stationary states. J Physics A 36: 631–641.

Gill AE (1982) Atmosphere-Ocean Dynamics. Academic Press, New York, NY.

Goody R (2000) Sources and sinks of climate entropy. Q J R Meteorol Soc 126: 1953–1970.

Grassl H (1981) The climate at maximum-entropy production by meridional atmospheric and oceanic heat fluxes. Q J R Meteorol Soc 107: 153–166.

Held IM, Hou AY (1980) Non-linear axially symmetric circulations in a nearly inviscid atmosphere. J Atmos Sci 37: 515–533.

Held IM, Suarez MJ (1994) A proposal for the intercomparison of the dynamical cores of atmospheric general circulation models. Bull Am Met Soc 75: 1825–1830.

James IN, Gray LJ (1986) Concerning the effect of surface drag on the circulation of a planetary atmosphere. Q J R Meteorol Soc 112: 1231–1250.

Kleidon A (2004) Beyond Gaia: Thermodynamics of Life and Earth system functioning. Clin Change, in press.

Kleidon A, Fraedrich K, Kunz T, Lunkeit F (2003) The atmospheric circulation and states of maximum entropy production. Geophys Res Lett 30: 2223.

Lorenz EN (1960) Generation of available potential energy and the intensity of the general circulation, in Dynamics of Climate. in: Pfeffer RL (ed), Dynamics of climate, Pergamon Press, Oxford, pp 86–92.

Lorenz RD (2001) Titan, Mars and Earth: Entropy production by latitudinal heat transport. Geophys Res Lett 28: 3169–3169.

Marshall J, Hill C, Perelman L, Adcroft A (1997a) Hydrostatic, quasi-hydrostatic, and non-hydrostatic ocean modeling. J Geophys Res 102(C3): 5733–5752.

Marshall J, Adcroft A, Hill C, Perelman L, Heisey C (1997b) A finite-volume, incompressible Navier Stokes model for studies of the ocean on parallel computers. J Geophys Res 102(C3): 5753–5766.

Nicolis G, Nicolis C (1980) On the entropy balance of the earth-atmosphere system. Q J R Meteorol Soc 106: 691–706.

Ozawa H, Ohmura A (1997) Thermodynamics of a global-mean state of the atmosphere – A state of maximum entropy increase. J Clim 10: 441–445.

Ozawa H, Shimokawa S, Sakuma H (2001) Thermodynamics of fluid turbulence: A unified approach to the maximum transport properties. Phys Rev E 64: 026303.

Ozawa H, Ohmura A, Lorenz RD, Pujol T (2003) The second law of thermodynamics and the global climate system – A review of the maximum entropy production principle. Rev Geophys 41: 1018.

Paltridge GW (1975) Global dynamics and climate – System of minimum entropy exchange. Q J R Meteorol Soc 101(429): 475–484.

Paltridge GW (1978) Steady-state format of global climate. Q J R Meteorol Soc 104: 927–945.

Pauluis O, Held IM (2002a) Entropy budget of an atmosphere in radiative-convective equilibrium. Part I: Maximum work and frictional dissipation. J Atmos Sci 59: 126–139.

Pauluis O, Held IM (2002b) Entropy budget of an atmosphere in radiative-convective equilibrium. Part II: Latent heat transport and moist processes. J Atmos Sci 59: 140–149.
Peixoto J, Oort A, Almeida M, Tome A (1991) Entropy budget of the atmosphere. J Geophys Res 96(D6): 10981–10988.
Pierrehumbert RT (2002) The hydrologic cycle in deep-time climate problems. Nature 419: 191–198.
Pujol T, Llebot JE (2000) Extremal climatic states simulated by a 2-dimensional model Part I: Sensitivity of the model and present state. Tellus 52A: 422–439.
Shimokawa S, Ozawa H (2002) On the thermodynamics of the oceanic general circulation: Irreversible transition to a state with higher rate of entropy production. Q J R Meteorol Soc 128: 2115–2128.
Shutts GJ (1981) Maximum entropy production states in quasi-geostrophic dynamical models. Q J R Meteorol Soc 107: 503–520.

9 Water Vapor and Entropy Production in the Earth's Atmosphere

Olivier M. Pauluis

Courant Institute of Mathematical Sciences, New York University, Warren Weaver Hall, 251 Mercer St., New York, NY 10012-1185, USA

Summary. This chapter investigates the production of entropy by moist processes in the Earth's climate system. The atmospheric circulation can be considered as a dehumidifier and heat engine operating in parallel: the dehumidification process reduces the system's overall efficiency far below the Carnot limit. In idealized simulations, moist processes are found to be a much stronger source of entropy production than is frictional dissipation. These results are then discussed in the context of the Earth's entropy production budget.

9.1 Introduction

The Earth is heated by the incoming solar radiation and is cooled by the emission of infrared radiation. Over long time scales, the Earth's atmosphere can be considered to be in statistical equilibrium, and its total energy and entropy remain approximately constant. This requires that the incoming and outgoing radiative fluxes balance each other (neglecting the small geothermal heat flux). As shortwave radiation carries less entropy than longwave radiation, the combination of incoming shortwave and outgoing longwave radiative fluxes results in a net export of entropy out of the atmosphere. This destabilizes the atmosphere and prevents it from reaching a global thermodynamic equilibrium. The loss of entropy is required to compensate for the net entropy production due to the various irreversible processes taking place within the Earth system.

Many processes contribute to this irreversible entropy production. The chapter focuses on the entropy production associated with the atmospheric circulation. Hence, although radiative processes are responsible for most of the entropy produced, radiative transfer bears little impact on the atmospheric flow. The discussion of the entropy budget of the atmospheric circulation can be greatly simplified by treating radiation as an external heat source or sink. The corresponding entropy source or sink is then equal to the total radiative energy absorbed or emitted, divided by the absolute temperature at which the absorption or emission takes place. The entropy budget of the atmospheric circulation takes the form:

$$Q_{surf}/T_{surf} + Q_{rad}/T_{rad} + \Delta S_{irr} = 0 \qquad (9.1)$$

Here, Q_{surf} is the heat flux at the Earth's surface, Q_{rad} is the net radiative cooling of the atmosphere, T_{surf} and T_{rad} are respectively the average temperature at which the heating and cooling occur, and ΔS_{irr} is the total irreversible entropy production by the atmospheric circulation. Energy conservation requires that heat sources and sinks balance each other:

$$Q_{surf} + Q_{rad} = 0 \tag{9.2}$$

As radiation is absorbed at a higher temperature than it is emitted, it acts as a net sink of entropy, which balances the total irreversible entropy production:

$$\Delta S_{irr} = \frac{T_{surf} - T_{rad}}{T_{surf} T_{rad}} Q_{surf} \tag{9.3}$$

The radiative forcing of the atmosphere determines the total irreversible entropy production, but does not provide any indications as to the nature of the irreversibilities. A wide variety of atmospheric phenomena can contribute to the total entropy production. For atmospheric motions, the irreversibility is associated at the microphysical level with either diffusion of heat, diffusion of water vapor, irreversible phase transitions, or frictional dissipation.

One of the main motivations for studying the entropy production budget lies in its connection to the generation and dissipation of kinetic energy. The maintenance of the atmospheric circulation requires mechanical energy to be continuously generated to compensate for the loss due to frictional dissipation (see also Ito and Kleidon, this volume). It is usually argued that the atmosphere acts, at least, partially as a heat engine: as warm air rises and expands and cold air sinks and is compressed, the atmospheric circulation produces a net mechanical work by transporting heat from a warm source at the Earth surface to the cold sink within the troposphere. The entropy budget provides an important constraint on how much work is produced by such an atmospheric heat engine. First, the irreversible entropy production by frictional dissipation is given by

$$\Delta S_{irr} = D/T_D \tag{9.4}$$

where D is the dissipation rate of kinetic energy, and T_D is the temperature at which the dissipation occurs. The maximum work W_{max} is defined as the maximum amount of kinetic energy that would be produced and dissipated in the absence of any irreversible source of entropy other than dissipation:

$$W_{max} = T_D \, \Delta S_{irr} = T_D \frac{T_{surf} - T_{rad}}{T_{surf} T_{rad}} Q_{surf} \tag{9.5}$$

If frictional dissipation occurs at the same temperature as the cooling, this expression would be equal to the work done by a Carnot cycle. As long as the internal variations of temperature are small in comparison with the absolute temperature, W_{max} can be approximated by the work performed by a Carnot cycle.

This maximum work provides an upper bound on the work done by the atmospheric circulation. If irreversible processes other than frictional dissipation result in an entropy production $\Delta S_{irr,nf}$, the actual work is smaller than this maximum, with

$$W = W_{max} - T_D \, \Delta S_{irr,nf} \tag{9.6}$$

A fundamental question is whether irreversible processes significantly reduce the amount of kinetic energy generated and dissipated by the circulation.

Pauluis and Held (2002a) investigate this issue by comparing the behavior of dry and moist convection. In the absence of water vapor, they find that the frictional dissipation is indeed the main irreversible source of entropy. However, in a moist atmosphere, diffusion of water vapor and irreversible phase changes account for a large portion, roughly two thirds, of the total entropy production. This is a consequence of moist convection acting not only as a heat engine but also as a dehumidifier that continuously removes water vapor through condensation and precipitation. This dehumidification is compensated for by evaporation at the Earth's surface. However, once the outflow of convective towers subsides and becomes unsaturated, any addition of water vapor involves either diffusion of water vapor or/and irreversible evaporation: the moistening of unsaturated air is inherently irreversible. The entropy production due to diffusion of water vapor and irreversible phase changes can thus be viewed as a measure of how much dehumidification and re-moistening occurs within the atmosphere.

As the total irreversible entropy production by the atmospheric circulation is constrained by the radiative forcing, all irreversible processes are in competition with one another in terms of their entropy production. This key aspect of the entropy budget implies that the extent to which the atmosphere behaves as a dehumidifier limits its ability to act as a heat engine. Any increase in entropy production due to diffusion of water vapor implies a reduction of the amount of kinetic energy available to drive the atmospheric circulation. Hence, in order to build a theory for the intensity of the atmospheric circulation based on the entropy budget, one must assess the role of moist processes.

This chapter focuses primarily on the contribution of the hydrological cycle to the Earth's overall entropy production. In the next section, three different idealized cycles are introduced to illustrate how water vapor can reduce the mechanical efficiency of a heat engine. These findings are related in Sect. 3 to the notion of the atmospheric dehumidifier and to the idea that the hydrological cycle can limit the strength of the atmospheric circulation. In Sect. 4, the importance of precipitation as a dissipative process is discussed. These findings are then extended to a discussion of the entropy budget of the Earth's atmosphere in Sect. 5.

9.2 Idealized Cycles

Three idealized cycles shows how condensation and evaporation can affect the energetics of the atmosphere. Cycle A illustrates the case of a pure dehumidifier, in which a parcel of moist air transports heat from a warm source to a cold sink without performing any mechanical work on its environment. Cycle B shows that atmospheric dehumidification is closely associated with expansion of water vapor. Cycle C looks at the relative importance of dry air and water vapor for the production of mechanical work. These cycles are schematically shown in Fig. 9.1 and described in more detail in the following.

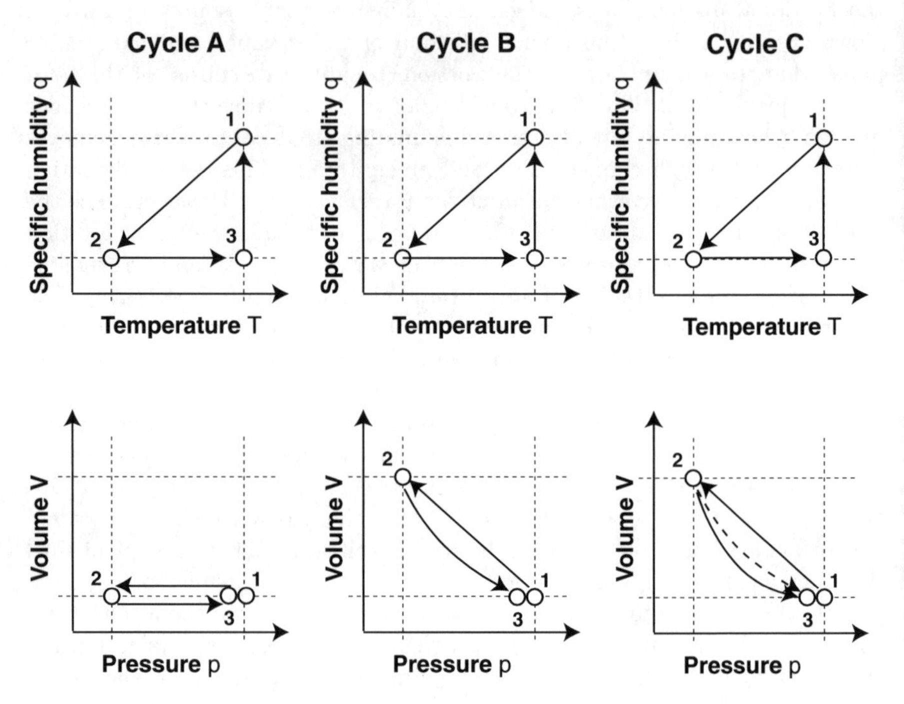

Fig. 9.1. Schematic diagram of three conceptual cycles leading to entropy production associated with atmospheric moisture and convection. The *dashed line* in the lower right diagram shows the corresponding trajectory along the moist adiabat of cycle B. See text for details

9.2.1 Cycle A: Pure Dehumidifier

Cycle A is a pure dehumidifier. A parcel of saturated moist air at temperature T_1 and mixing ratio q_1 is cooled to a temperature T_2. As water vapor condenses, the water vapor mixing ratio is reduced to q_2. In the second leg of the cycle, the parcel is heated back to the initial temperature $T_3 = T_1$, without allowing any of the condensed water to reevaporate, so that the mixing ratio remains constant $q_3 = q_2$. Finally, the cycle is closed by evaporating water at constant temperature T_1 up to the original mixing ratio q_1. These transformations take place at constant volume, so the system does not exert any work on the environment. Energy conservation implies that the cooling during the stage 1–2 must balance the sum of the sensible and latent heating associated with the transformation 2–3 and 3–1.

The only irreversibility in this system is the evaporation of water vapor into unsaturated air during the step 3–1. The entropy production due to the evaporation of an infinitesimal quantity dq of water vapor is $-R\,dq \ln e/e_s$, where R is the gas constant for water vapor, e is the partial pressure of water vapor and e_s is the saturation vapor pressure. The irreversible entropy production in cycle A is thus

$$\begin{aligned} \Delta S_{irr} &= \int_{q_3}^{q_1} \left(-R \ln \frac{q}{q_1} \right) dq \\ &= R(q_1 - q_3) + R q_3 \ln \frac{q_3}{q_1} . \end{aligned} \tag{9.7}$$

This irreversible entropy production is balanced by a net export due to external heat sources and sinks. The entropy export associated with the cooling during the first stage of the cycle is characterized by an effective temperature T_{out} given by

$$T_{out}^{-1} = \int_{q_3}^{q_1} \frac{1}{T} dq . \tag{9.8}$$

Since this temperature is lower than the temperature $T_{in} = T_1$ at which evaporation occurs, heat is transported from a warm source (T_{in}) to a colder sink (T_{out}). A calculation of this entropy export, using the Clausius-Clapeyron relationship, yields

$$\begin{aligned} \frac{Q_{in}}{T_{in}} + \frac{Q_{out}}{T_{out}} &= \int_{q_3}^{q_1} \left(\frac{1}{T_1} - \frac{1}{T} \right) L dq \\ &= \int_{q_3}^{q_1} R \ln \frac{q}{q_1} dq = -\Delta S_{irr} \end{aligned} \tag{9.9}$$

with L being the latent heat of vaporization. Comparing (9.7) and (9.9) shows that the entropy production due to the irreversible evaporation into unsaturated air is balanced by a net export due to the external heat sources.

9.2.2 Cycle B: Atmospheric Dehumidifier and Water Vapor Expansion

In the pure dehumidifier, all transformations take place at constant volume and no work is exerted on the environment. In the atmosphere however, expansion and compression of air parcels are a key aspect of the atmospheric circulation. We construct cycle B to investigate how water vapor expansion modifies the behavior of a dehumidifier.

A saturated parcel of moist air at stage 1 expands adiabatically and reversibly from the surface pressure p_1 to a pressure p_2. We use $T_{ad}(p_d)$ to describe the temperature of the air parcel during its ascent as a function of the partial pressure of dry air $p_d = p - e$. In the second leg of the cycle, the parcel is compressed and cooled back to its initial temperature and pressure, without allowing for any re-evaporation of the condensed water. The compression follows the same 'temperature path' $T_{ad}(p_d)$ as the expansion. In the third stage, the parcel is brought back to saturation at constant temperature and volume (and therefore constant p_d).

This cycle exerts an amount of work W_B on its environment, with

$$W_B = \oint p\,d\alpha = \oint (p_d + e) d\alpha = \oint e\,d\alpha\,, \tag{9.10}$$

where $d\alpha$ is an infinitesimal change in the specific volume of water vapor. The first part of the integral is the work performed by dry air and vanishes as the temperature is the same during the expansion and compression. The mechanical work produced by the cycle B is thus solely due to water vapor expansion.

Because the expansion 1–2 is adiabatic, there is no external heating during that portion of the cycle. The amount of cooling that takes place during the compression 2–3 is given by:

$$Q_{out} = \int_2^3 \delta Q = \int_2^3 dU + p d\alpha, \tag{9.11}$$

where dU is an infinitesimal change in internal energy. The external heat flux required for evaporation is

$$Q_{in} = L(q_1 - q_3) \tag{9.12}$$

The first law of thermodynamics implies that a net heating of the system is required to balance the amount of work W_B exerted on the environment:

$$Q_{in} + Q_{out} = W_B \tag{9.13}$$

The irreversible entropy production due to evaporation is equal to that in the pure dehumidifier (9.7). The entropy exports due to the cooling during the transition 2–3 is given by

$$\frac{Q_{out}}{T_{out}} = \int_2^3 \frac{\delta Q}{T} = \int_2^3 \frac{dU + p\, d\alpha}{T}. \tag{9.14}$$

The entropy budget requires that the entropy production ΔS_{irr} to be balanced by a net export of entropy through the differential heating:

$$Q_{in}/T_{in} + Q_{out}/T_{out} + \Delta S_{irr} = 0 \tag{9.15}$$

Eliminating Q_{out} in (9.15) by using (9.13) yields

$$W_B/Q_{in} = (T_{in} - T_{out})/T_{in} - T_{out}\, \Delta S_{irr}/Q_{in} \tag{9.16}$$

The left hand side is the mechanical efficiency of the system. The first term on the right-hand side is the efficiency of an equivalent Carnot cycle, and the second term is the loss resulting from irreversible evaporation.

A fundamental question here is that of the relative importance of the expansion work by water vapor and of the entropy production by moist processes. The efficiency of the cycle, defined as the ratio W_B/Q_{in}, is computed for various values of the initial temperature T_1, with $p_1 = 1000\,\text{mb}$, and $p_2 \approx 0\,\text{mb}$. The temperature of the moist adiabat drops very quickly, and there is little water vapor remaining above 200 mb, so the actual value of p_2 has little impact on the calculation as long as it is small enough. For a surface temperature corresponding to the current atmosphere $T_1 \approx 290\,\text{K}$, the efficiency of the cycle is about one quarter of the efficiency of the equivalent Carnot cycle, and both efficiencies increase with temperature.

The mechanical efficiency of the cycle is quite sensitive to the surface temperature T_1. Assuming $e \approx e_0\,(\alpha/\alpha_0)^{-\gamma}$, the work can be approximated by

$$W_B \approx \int_{\alpha_0}^{\alpha} e_0 \left(\frac{\alpha}{\alpha_0}\right)^{-\gamma} d\alpha \approx \frac{e_0 \alpha_0}{\gamma - 1} \approx q_1 \frac{RT}{\gamma - 1}. \tag{9.17}$$

Comparing this with the expression for the entropy production (9.7) yields

$$W_B/(W_B + T_{out}\, \Delta S_{irr}) \approx 1/\gamma \tag{9.18}$$

The parameter γ is approximately equal to the ratio between the scale height of density and the scale height for saturation water vapor pressure. For atmospheric conditions, these are approximately equal to 10,000 m and 2,500 m, so $\gamma \approx 4$ and the efficiency of the cycle is about one quarter of the efficiency of a Carnot cycle. When the surface temperature increases, the lapse rate of the moist adiabat decreases. This leads to a larger scale height for the saturation water vapor (and thus a smaller γ), and the efficiency of the cycle increases, as shown in Fig. 9.2.

Notice that the increased efficiency is not the result of a decrease in entropy production by moist processes, but rather is due to an increase of the

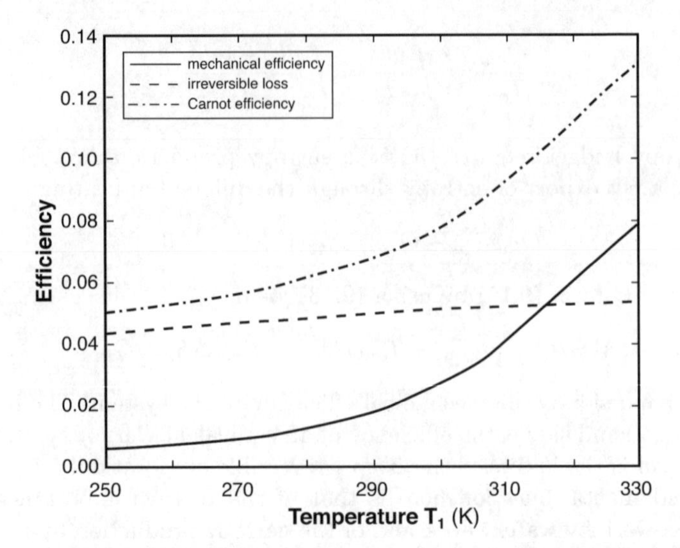

Fig. 9.2. Mechanical efficiency W_B/Q_{in} (*solid line*), irreversible loss $T_{out}\,\Delta S_{irr}/Q_{in}$ (*dashed line*) and Carnot efficiency $(T_{in} - T_{out})/T_{in}$ for cycle B. These three terms corresponds to the balance in (9.16). See text for details

temperature difference between the heat source and sink. Indeed, comparing (9.7) and (9.12) shows that for $q_3 \approx 0$, the reduction of efficiency due to irreversible evaporation is approximately constant with

$$T_{out}\,\Delta S_{irr}/Q_{in} = RT/L \approx 0.05 \tag{9.19}$$

The difference between the actual efficiency and that of a Carnot cycle is approximately equal to 0.05, and does not depend significantly on the temperature of the system.

9.2.3 Cycle C: Sensible Heat Transport

In the previous cycle, dry air contributes neither to heat transport nor to mechanical work. In the atmosphere, however, the combination of warm air expansion and cold air compression plays an important part in both. Cycle C is designed to include this contribution. Cycle C is identical to cycle B except that the temperature of the air during compression is given by $T_C(p_d)$.

Cycle C is equivalent to a combination of cycle B and a dry subcycle in which an air parcel goes from 3 to 2 following a temperature path given by $T_{ad}(p_d)$ then returns to 3 with a temperature $T_C(p_d)$. This subcycle does not involve any phase transition and can be considered as acting as a perfect heat engine. Assuming $T_C(p_d) = T_{ad}(p_d) - \Delta T$, the work done by cycle C can be approximated by

$$W_C = W_B + \frac{\Delta T}{T_{out,B}}\left|Q_{out,B}\right|, \tag{9.20}$$

where W_B, $Q_{out,B}$, $T_{out,B}$ are respectively the work, cooling, and effective cooling temperature obtained for cycle B.

9.3 Dehumidifier Versus Heat Engine

These three cycles emphasize the fact that the mechanical work produced by transport of latent heat from a warm source to a cold sink is smaller than the theoretical maximum of a Carnot cycle. This reduction of work can be viewed as a direct consequence of the atmospheric dehumidification.

Imagine an ideal membrane that is impermeable to water vapor but permeable to dry air. Such a membrane makes it possible to *reversibly* condense all water vapor in a parcel of moist air at constant temperature by slowly pushing down the membrane and reducing the volume occupied by the water vapor, so as to keep the vapor pressure constant. The work required to push down the membrane is $\alpha\, e = q_1\, RT$, and the cooling necessary to keep the temperature constant is $q_1\,(L + RT)$. Reversible evaporation is also possible by slowly lifting the membrane and would exert work on the environment. In the absence of the membrane, evaporation is irreversible. In this case, the required heating is only $q_1\, L$, as no work is performed during evaporation. The irreversible entropy production $q_1\, R = \alpha\, e/T$ is proportional to the amount of work that would have been produced by a reversible transformation but was "lost" due to the irreversible nature of evaporation.

In the atmosphere, the absence of an ideal membrane also makes it impossible to remove water vapor by compressing it at constant temperature. Instead, condensation can only occur after the saturation vapor pressure has been lowered by reducing the temperature. In statistical equilibrium, condensation must take place at a lower average temperature than evaporation. Atmospheric dehumidification is thus associated with latent heat transport from a warm source to a cold sink. As illustrated by the cycles above, this latent heat transport produces less work on its environment than a corresponding Carnot cycle. This reduction of the mechanical output can be interpreted as the amount of work that must be produced internally in order to compress the water vapor into a liquid form, and that is being lost during the subsequent irreversible evaporation.

A second aspect of the atmospheric dehumidification is that it is closely tied to air ascent and adiabatic expansion. It is therefore closely associated with water vapor expansion. Results from cycle B indicate that the amount of work performed by water vapor expansion is only a fraction of the work by a Carnot cycle. Pauluis and Held (2002b) generalize this result to an atmosphere in statistical equilibrium. They find that, for tropical conditions, the mechanical efficiency of the latent heat transport is about one quarter of the efficiency of perfect heat engine. For Earth-like condition, the latent heat transport acts primarily as a dehumidifier.

This statement is only valid for conditions similar to that of the Earth atmosphere. Figure 9.2 shows that the contribution of water vapor expansion increases with surface temperature. The ratio of water vapor expansion versus dehumidification is highly sensitive to the atmospheric lapse rate, and hence to the initial water vapor mixing ratio. If water vapor were the main atmospheric constituent, the temperature profile of a moist adiabat would be determined by the saturation curve so that the total pressure would be approximately equal to the saturation vapor pressure $p \approx e_s(T)$. Latent heat transport would be primarily associated with water vapor expansion, and irreversible entropy production due to evaporation would be negligible. The Earth's atmosphere is far from this situation, but this is a reasonable scenario for Mars in which the primary atmospheric constituent, carbon dioxide, can condense (see also Lorenz, this volume).

Atmospheric dehumidification implies that the mechanical output of latent heat transport is smaller than the Carnot efficiency. However, this argument does not affect the portion of the heat transport due to sensible heat. As illustrated by cycle C, the overall efficiency of the atmospheric circulation is highly sensitive to the partitioning of the total heat transport into sensible and latent heat. When latent heat transport is dominant, entropy production due to irreversible evaporation is large and the cycle produces much less work than the corresponding Carnot cycle. Furthermore, water vapor expansion is responsible for most of the production of mechanical work. Conversely, when latent heat is small in comparison to the sensible heat transport, moist processes only play a small role and the mechanical efficiency of the atmosphere should be close to that of a Carnot cycle.

9.4 Frictional Dissipation in Falling Precipitation

The two previous sections focus on how the hydrological cycle reduces the generation of kinetic energy in the atmosphere. However, the hydrological cycle also plays an important role in how kinetic energy is dissipated.

Atmospheric motions generate mechanical energy at scales much larger than those at which viscosity can efficiently remove kinetic energy. It is usually argued that under these conditions dissipation occurs as the end result of a turbulent energy cascade. However, the hydrological cycle offers a simple way to by-pass the turbulent cascade. Indeed, hydrometeors generate a microscopic shear zone that acts to slow down their fall. The shear extracts mechanical energy from the hydrometeors and transfers it to the microscopic flow, where it is dissipated. For a hydrometeor falling at its terminal velocity v_T, the dissipation rate is given by $M\,g\,v_T$ where M is the mass of the hydrometeor and g is the acceleration by gravitation. For precipitation through a stationary atmosphere, the dissipation rate is thus equal to the rate of change of the geopotential energy of the hydrometeors.

As the mechanical work required to lift water is continuously dissipated during precipitation, the amount of kinetic energy available to drive a circulation on larger scales is reduced. Pauluis et al. (2000) find that, in their numerical simulations, more energy is dissipated by falling precipitation than is used to generate convective motions. They also estimate that the frictional dissipation in falling precipitation to be of the order of 2–4 $\mathrm{W\,m^{-2}}$ averaged over the tropical atmosphere. This dissipation rate depends on several mechanisms, such as entrainment, re-evaporation of precipitation, and water loading, and this estimate should be viewed as tentative. Nevertheless, precipitation is most likely an important form of dissipation in the Earth's atmosphere, and quite possibly in others (Lorenz and Rennò, 2002).

9.5 Entropy Budget of the Earth's Atmosphere

Pauluis and Held (2002a) analyze the entropy budget of moist convection in radiative-convective equilibrium. They use a high-resolution numerical model to simulate moist convection in statistical equilibrium with radiative and surface forcings. They find that diffusion of water vapor and irreversible phase changes account for about two thirds of the total entropy production, while frictional dissipation in falling precipitation account for another 20% . The amount of kinetic energy generated at convective scales and dissipated through a turbulent energy cascade accounted for less than 10% of the entropy production. This leads them to characterize moist convection as behaving more as an atmospheric dehumidifier than as a heat engine.

Pauluis and Held (2002a) focus on the behavior of an atmosphere in radiative-convective equilibrium, in which the external forcings (radiative and surface fluxes) are horizontally uniform. Hence, these simulations are not meant to be representative of the whole atmosphere but rather focus on the behavior of tropical convection. The absence of rotation allows for gravity waves to quickly spread horizontal temperature fluctuations over large distance. As the horizontal temperature variations are small, the sensible heat transport is weak. Latent heat transport dominates the vertical heat transport, and moist processes dominate the entropy budget. The Earth's rotation and the longitudinal variations of the radiative forcing result in an Equator-to-Pole temperature difference much larger than the horizontal temperature fluctuations within tropical convection. The contribution of sensible heat transport is thus much stronger for the heat transport by the large-scale atmospheric circulation than it is for tropical convection. This implies a higher mechanical efficiency and smaller contribution of moist processes. Nevertheless, given the large contribution of deep tropical convection in the vertical heat transport, it is likely that the hydrological cycle still plays a large role in the entropy budget.

Quantitative estimates of entropy production in the Earth's atmosphere are difficult to establish with high accuracy, as they depend on global measurements of small-scale processes. Peixoto et al. (1991) discuss the entropy budget of the atmosphere based on a simple box model (see also Fig. 1.2). However, they treat moist processes as an external heat source, and, by doing so, they neglect the irreversibility associated with the hydrological cycle. Goody (2000) revisits their analysis by improving the estimate of frictional dissipation and considering more carefully the impact of moist processes. The introductory chapter to this book also provides some discussion.

The same box model is used both by Peixoto et al. (1991) and by Goody (2000). It considers a surface latent heat flux of $79\,\mathrm{W\,m^{-2}}$ and a surface sensible heat flux of $20.4\,\mathrm{W\,m^{-2}}$ occurring at an average surface temperature of 288 K. This is balanced by a net radiative cooling of the troposphere of $99.4\,\mathrm{W\,m^{-2}}$ at an average temperature 255 K. This yields a net export of entropy of $45\,\mathrm{mW\,m^{-2}\,K^{-1}}$. If the average temperature at which friction occurs is 280 K, this corresponds to a maximum work of $12.5\,\mathrm{W\,m^{-2}}$.

Peixoto et al. (1991) estimate a dissipation rate for large-scale atmospheric motions of about $1.9\,\mathrm{W\,m^{-2}}$ yielding an entropy source of $7\,\mathrm{mW\,m^{-2}\,K^{-1}}$. Goody (2000) considers additional dissipation mechanisms and finds an entropy production due to frictional dissipation of $11.3\,\mathrm{mW\,m^{-2}\,K^{-1}}$. These do not include the contribution of falling precipitation. The entropy source for dry convection is estimated at $2\,\mathrm{mW\,m^{-2}\,K^{-1}}$ and should be interpreted as the result of frictional dissipation in shallow non-precipitating convection.

As they treat condensation as an external heat source, Peixoto et al. (1991) do not provide any entropy source associated with moist processes. They do however estimate the average condensation temperature at 266 K. If one assumes that the mechanical efficiency of latent heat transport is one quarter that of a Carnot cycle in agreement with Pauluis and Held (2002b), the transport of latent heat would be associated with work by water vapor expansion of $1.6\,\mathrm{W\,m^{-2}}$, and an irreversible entropy production due to diffusion of water vapor and irreversible phase changes of $17\,\mathrm{mW\,m^{-2}\,K^{-1}}$. Goody (2000) uses a slightly different method and estimates this source of entropy production to be only $13.3\,\mathrm{mW\,m^{-2}\,K^{-1}}$. The small difference is probably due to the fact that Goody's method uses an implicit value for the average condensation temperature that is higher than 266 K. Finally, Goody (2000) estimates that dissipation by falling precipitation generates about $5.5\,\mathrm{mW\,m^{-2}\,K^{-1}}$ of entropy. While this number is arbitrary, it lies in the range expected from the arguments of Pauluis et al. (2000).

Together, these estimates indicate that frictional dissipation of atmospheric motions accounts for about 30% of the total entropy production by the atmospheric circulation, that frictional dissipation in falling precipitation explains an additional 12% , and that the phase changes and diffusion of water vapor are responsible for about 40% . The remaining 20% is indicative of the fairly large uncertainty of these estimates of the entropy production. When considering the atmosphere as a whole, the hydrological cycle is directly responsible for roughly half the entropy production by the atmospheric circulation.

References

Goody R (2000) Sources and sinks of climate entropy. Q J Roy Meteorol Soc 126: 1953–1970.

Lorenz RD, Rennò NO (2002) Work Output of Planetary Atmospheric Engines: Dissipation in Clouds and Rain. Geophys Res Lett 29(2): 1023.

Pauluis OM, Held IM (2002a) Entropy budget of an atmosphere in radiative-convective equilibrium. Part I: maximum work and frictional dissipation. J Atmos Sci 59: 125–139.

Pauluis OM, Held IM (2002b) Entropy budget of an atmosphere in radiative-convective equilibrium. Part II: Latent heat transport and moist processes. J Atmos Sci 59: 140–149.

Pauluis OM, Balaji V, Held IM (2000) Frictional dissipation in a precipitating atmosphere. J Atmos Sci 57: 987–994.

Peixoto JP, Oort AH, de Almeida M, Tomé A (1991) Entropy budget of the atmosphere. J Geophys Res 96(D6): 10981–10988.

References

[illegible] (1996) [illegible]

[illegible] (2000) [illegible] Elementary [illegible]

[illegible]

[illegible] (1982) [illegible]

[illegible] and [illegible] (2001) [illegible]

[illegible]

[illegible] [illegible] (1999) [illegible]

10 Thermodynamics of the Ocean Circulation: A Global Perspective on the Ocean System and Living Systems

Shinya Shimokawa[1] and Hisashi Ozawa[2]

[1] National Research Institute for Earth Science and Disaster Prevention, Tsukuba 305-0006, Japan
[2] Institute for Global Change Research, Frontier Research System for Global Change, Yokohama 236-0001, Japan

Summary. In this chapter, we investigate thermodynamics in a global-scale open ocean circulation and discuss the physical properties of "living systems", that is, individual organisms, by analogy to the behavior of the ocean system. Despite the fact that the ocean system has long been examined from a dynamic point of view, its thermodynamic aspects remain to be explored. We show a quantitative method that expresses the rate of entropy production in an open dissipative system that exchanges heat and matter with its surrounding system. This method is applied to an ocean circulation model, and the rate of entropy production is examined in relation to the dynamic behavior of the system. Multiple steady states can exist under the same set of boundary conditions, and the state can be shifted by applying perturbations at the surface boundary. The perturbations tend to shift the system to a state of higher entropy production, except when a perturbation destroys the initial circulation completely. This result supports the hypothesis that a nonlinear dynamic system tends to move to a state with higher entropy production by producing an active circulation in the system when triggered by perturbations. When such a system is subject to random perturbations for a certain period of time, the most probable state to result will be the one with the maximum entropy production. The entropy produced in a steady-state dissipative system is discharged into the surrounding system through boundary fluxes of heat and matter, thereby contributing to the entropy in the surrounding system. Finally, an analogy is suggested between the ocean system and a living system, in which a highly organized circulatory structure of fluids has evolved from a less organized primeval one, thereby producing entropy in the surrounding system at an increased rate.

10.1 Introduction

The world's oceans can be seen as an open dissipative system connected with its surroundings by the exchange fluxes of heat and freshwater. The surrounding system consists of the other components of the Earth system, specifically the atmosphere, the cryosphere, and the land surface, and ultimately space. Because of the curvature of the Earth's surface and the inclination of its rotation axis relative to the Sun, the ocean receives a net gain of heat from solar radiation in the equatorial region, and high evaporation rates lead to the removal of freshwater from the ocean surface, resulting in an apparent positive salt flux. In polar regions, the ocean looses heat through long-wave emissions into space, and high precipitation, snow and ice melt result in a positive freshwater flux at the ocean surface (an apparent negative salt flux). Thus, in general, there are net gains of heat and salt in the equatorial region, and net losses of heat and salt in the polar regions. (Strictly speaking, there is a slight loss of salt in the central equatorial region, but the amount is much less significant than those in the polar regions.) The net flux imbalance brings about an inhomogeneous distribution of temperature and salinity at the surface, that is, a warm, saline surface ocean in equatorial region and a cold, less saline surface ocean in polar regions. This inhomogeneity creates a circulation, which in turn reduces this inhomogeneity. In this respect, the formation of a circulatory system can be regarded as a process leading to the final equilibrium of the whole system: the ocean system and its surroundings. In this process, the rate of approaching the equilibrium, that is, the rate of entropy production by the oceanic circulation, appears to be an important factor.

It has been suggested that the rate of entropy production in an open dissipative system is related to the stability of the system. Two ideas have been proposed to explain the relationship between the rate of entropy production and the stability of a system. Prigogine (1955) showed that a linear irreversible process becomes stabilized in a state with minimum entropy production (the principle of minimum entropy production, see also the introduction, Kleidon and Lorenz, this volume). Glansdorff and Prigogine (1964) suggested that this principle may also be valid for nonlinear processes. Conversely, Sawada (1981) stated that the entropy of thermal reservoirs connected via a nonlinear system will increase along an evolutionary path, favoring the maximum entropy production among manifold possible paths (the principle of maximum entropy production, or MEP, also Dewar 2003, and this volume). He regarded Prigogine's principle of minimum entropy production as a trivial one, valid only for systems close to linear and equilibrium regions (see also Ozawa et al. 2003). There is, in fact, some evidence to support the principle of maximum entropy production for highly nonlinear and nonequilibrium systems. For example, Paltridge (1975, 1978) suggested that the global distribution of the present climate is reproducible as a state with maximum entropy production by turbulent heat transport in the atmosphere and oceans. Several researchers have investigated his work and obtained similar results (Grassl, 1981; Noda and Tokioka, 1983; Ozawa and Ohmura, 1997). This principle may also be valid for other systems: for example, a Bénard-type convection system (Suzuki and Sawada, 1983), the atmospheres of Mars and Titan (Lorenz et al., 2001), the mantle–core system of the Earth (Vanyo and Paltridge, 1981), hydrodynamic pattern formation (Woo, 2002), crystal morphology transition (Hill, 1990; Martyushev et al., 2000), and bacterial metabolism (Forrest and Walker, 1964). Nevertheless, no study has yet been carried out to test the principle of maximum entropy production for the ocean system.

A living organism can also be seen as an open dissipative system (Schrödinger, 1944). Although several attempts have so far been made to link the activity of living systems to maximum entropy production (Lotka, 1922; Ulanowicz and Hannon, 1987; Sawada, 1994), the thermodynamic properties of living systems associated with their evolution remain to be explored (see also Catling, this volume, Kleidon and Fraedrich, this volume, Toniazzo et al., this volume).

In this chapter, we investigate the thermodynamics of a global-scale open ocean system. In particular, we focus on whether or not the interactions of the open ocean system with its surroundings are dominated by the principle of maximum entropy production. For this purpose, we present an equation that expresses the rate of entropy production in an open dissipative system that interacts with its surrounding system. By applying the equation to an oceanic general circulation model, we examine the entropy production in a steady state of the ocean system and the transition among multiple steady states (the existence of multiple steady states has long been a point of debate in physical oceanography, see e.g., Stommel 1961, Bryan 1986, Marotzke and Willebrand 1991 for details). In addition, we examine the evolution of the oceanic structure in the transition among multiple steady states. Finally, we discuss living systems and the applicability of MEP by analogy to the modeling results obtained for the ocean circulation.

10.2 Calculation of Entropy Production

The rate of entropy production in an open ocean system can be expressed as (Shimokawa and Ozawa, 2001):

$$\dot{S} = \int \frac{\rho c}{T} \frac{\partial T}{\partial t} dV + \int \frac{F_h}{T} dA - \alpha k \int \frac{\partial C}{\partial t} \ln C dV - \alpha k \int F_s \ln C dA \,, \tag{10.1}$$

where ρ is the density, c is the specific heat at constant volume, T is the temperature, $\alpha = 2$ is van't Hoff's factor representing the dissociation of salt into separate ions (Na^+ and Cl^-), k is the Boltzmann constant, C is the concentration of salt per unit volume of sea water, F_h and F_s are the heat and salt fluxes per unit surface area, defined as positive outward (upward), respectively, dV is a small volume element (volume integrations are taken over the whole ocean volume), and dA is a small surface element (surface integrations are taken over the whole ocean surface). The first term on the right-hand side represents the rate of entropy increase in the ocean system due to heat transport, and the second term represents that in the surrounding system. The third term represents the rate of entropy increase in the ocean system due to salt transport, and the fourth term represents that in the surrounding system. Overall, (10.1) represents the rate of entropy increase of the whole system, that is, the rate of entropy production due to irreversible processes associated with the oceanic circulation. This equation is applicable to a large-scale circulation model whose scale of resolution is very much coarser than the dissipation scale, because it does not include a microscopic representation of the dissipation process.

10.3 Model Description and Experimental Method

The numerical model used in this study is Pacanowski's (1995) version of the Geophysical Fluid Dynamics Modular Ocean Model. The model domain is a rectangular basin with a cyclic path, representing an idealized Atlantic Ocean (Fig. 10.1a). The horizontal grid spacing is 4°. The depth of the ocean is 4500 m with 12 vertical levels. Sub-grid-scale eddy transport coefficients are as follows: horizontal diffusivity $10^3\,\mathrm{m^2\,s^{-1}}$, horizontal viscosity $8 \times 10^5\,\mathrm{m^2\,s^{-1}}$, vertical diffusivity $10^{-4}\,\mathrm{m^2\,s^{-1}}$, and vertical viscosity $2 \times 10^{-3}\,\mathrm{m^2\,s^{-1}}$.

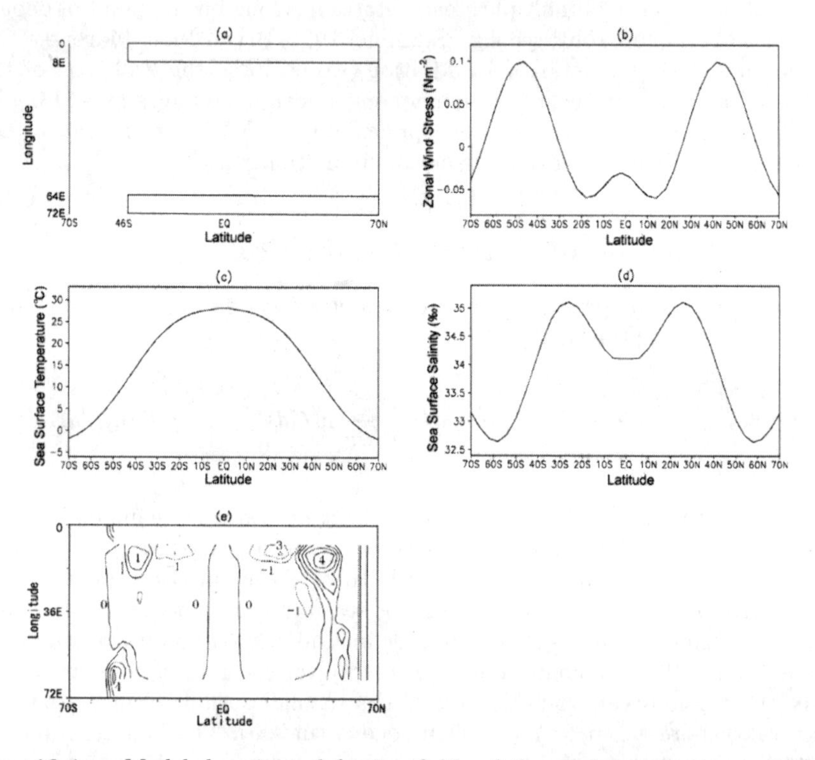

Fig. 10.1. **a** Model domain and forcing fields of the model as functions of latitude, **b** forced zonal wind stress ($\mathrm{N\,m^{-2}}$) defined as positive eastward, **c** prescribed sea surface temperature (°C), **d** prescribed sea surface salinity (‰), **e** freshwater flux diagnosed from steady state $\mathrm{N_{RBC}}$ (Fig. 10.2a). The contour line interval of freshwater flux is 1.0 m $\mathrm{year^{-1}}$

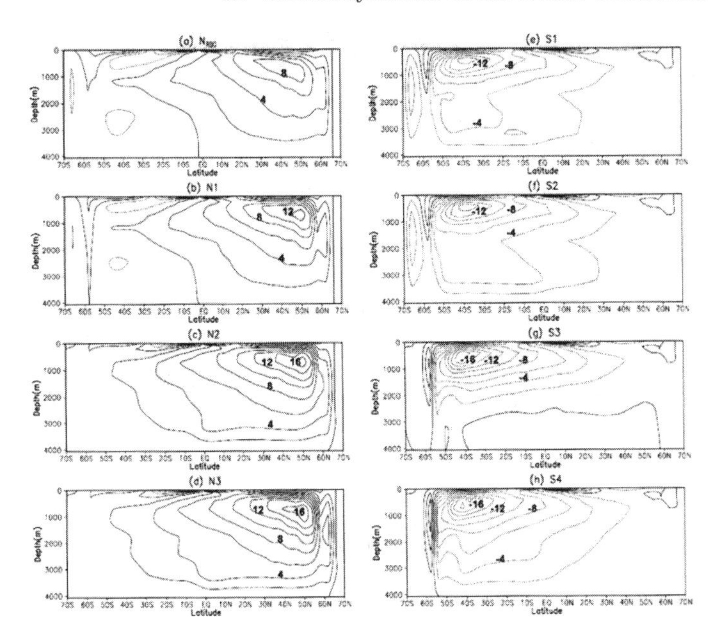

Fig. 10.2. All steady states obtained from this study. **a** N_{RBC}, **b** N1, **c** N2, **d** N3, **e** S1, **f** S2, **g** S3, **h** S4. Fields shown are zonally integrated meridional stream functions at year 5000 for (a) spin-up and at year 1500 for the transition experiments **b** r06, **c** r02, **d** r16, **e** r01, **f** r04, **g** r13, and **h** r14 (see also Fig. 10.4). The unit is SV ($= 10^6\,m^3\,s^{-1}$). The contour line interval is 2 SV. The capital letters "N" and "S" refer to northern and southern sinking, respectively. N_{RBC} is a unique solution under restoring boundary conditions

Our experiments consist of three phases (Shimokawa and Ozawa, 2002): (1) spin-up under restoring boundary conditions (Figs. 10.1b, c, d) for 5000 years, (2) integration under mixed boundary conditions (Figs. 10.1b, c, e) with a high-latitude salinity perturbation for 500 years, and (3) integration under mixed boundary conditions without perturbation for 1000 years. As a result of phase (1), the system reaches a statistically steady state with northern sinking circulation (N_{RBC}, Fig. 10.2a). In phase (2), the system moves to a state determined by the perturbation applied. In phase (3), the system is adjusted satisfactorily to the boundary condition without perturbation. Then, in some cases, the system returns to the initial state, while in other cases it does not, instead remaining in the state determined by the perturbation or moving to a different steady state which is independent of the perturbation. If a new steady state is obtained, phases (2) and (3) are repeated using the new steady state as the initial state. If a new steady state is not obtained, these procedures are repeated using a different salinity perturbation. As a result, a series of multiple steady states of thermohaline circulation under the same set of wind forcing and mixed boundary conditions are obtained. All the steady states obtained from this study are shown in Fig. 10.2. The standard salinity perturbation Δ used in this study is $2 \times 10^{-7}\,\mathrm{kg\,m^{-2}\ s^{-1}}$ (corresponding to a freshwater flux of $\approx -0.1\,\mathrm{m\,year^{-1}}$), which is usually applied north of 46° N and occasionally south of 46° S.

10.4 Entropy Production in a Steady State

Figure 10.3 shows the entropy increase rates due to the heat and salt transports calculated with (10.1) during the spin-up period. In the steady state after the year 4000, for both heat and salt transport, the entropy increase rates for the ocean system are zero, whereas those for the surrounding system show positive values, within the limits of the accuracy of our numerical model. The zero increase rates represent the fact that the ocean system is in

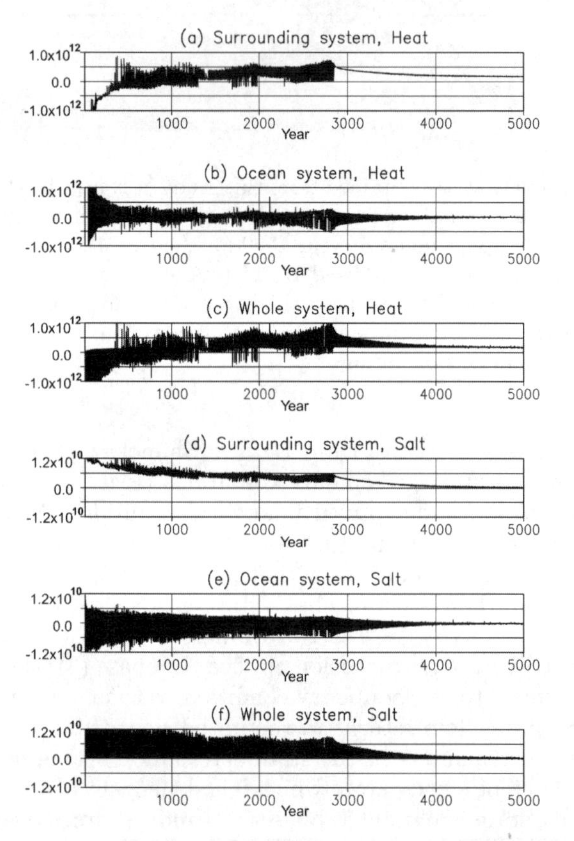

Fig. 10.3. Entropy increase rates calculated during the spin-up period for: **a** the surrounding system by heat transport, **b** the ocean system by heat transport, **c** the whole system [(a)+(b)] by heat transport, **d** the surrounding system by salt transport, **e** the ocean system by salt transport, and **f** the whole system [(d)+(e)] by salt transport

a steady state, so the distributions of temperature and salinity remain unchanged ($\partial T/\partial t = \partial C/\partial t = 0$); hence the volume integrals in (10.1) result in zero. In contrast, heat and salt are transported from equatorial (warm and salty) to polar (cold and less salty) regions by the steady-state circulation. These irreversible transports contribute to the entropy increase in the surrounding system through the surface integrals in (10.1). In this sense, the surrounding system is *not* in a steady-state, and its entropy is increasing slowly but steadily by the contribution of the oceanic circulation. The entropy increase rates in the surrounding system are $1.9 \times 10^{11}\,\mathrm{W\,K^{-1}}$ by the heat transport and $3.6 \times 10^{8}\,\mathrm{W\,K^{-1}}$ by the salt transport. The former is three orders of magnitude larger than the latter. Thus, the entropy increase is due primarily to heat transport, being consistent with an earlier estimate by Gregg (1984). In conclusion, the produced entropy in a steady-state ocean system is completely discharged into the surrounding system through the boundary fluxes of heat and matter, thereby contributing to the entropy in the surrounding system.

10.5 Entropy Production During Transition Among Multiple Steady States

Next, we show how the ocean circulation responds when the steady states are perturbed. The results of the transition experiments are summarized in Fig. 10.4, showing the relationship between the transitions among the multiple steady states and the associated rates of entropy production. Starting from S1 (Fig. 10.2e), the system moves to S2 (Fig. 10.2f) regardless of the sign of the perturbation (r04 and r05); whereas starting from S2, the system does not return to S1, but remains in the initial state (S2) regardless of the sign of the perturbation (r08 and r09). In addition, starting from S3 (Fig. 10.2g), the system moves to S4 (Fig. 10.2h) regardless of the sign of the perturbation (r14 and r15); whereas starting from S4, the system does not return to S3, but remains in the initial state (S4) regardless of the sign of the perturbation (r18 and r19). When these transitions occur (r04, r05, r14 and r15), the rates of entropy production in the final states are always higher than those in the initial states (see Fig. 10.4). These results show that the transition from a state with lower entropy production to a state with higher entropy production tends to occur, but the transition in the reverse direction does not occur, i.e., the transition is *irreversible* or *directional* in the direction of the increase in the rate of entropy production. Here, the term "irreversible" is used for cases in which the transition from state A to state B occurs when a finite perturbation X is applied to A, but the transition from B to A does not occur when a perturbation with the inverse sign and the same strength (−X) is applied to B. These irreversible transitions are consistent with the principle of maximum entropy production. On the other hand, for the transition from northern sinking to southern sinking, the rate of entropy production can decrease, such as r12 and r13. These results seem to contradict the principle of maximum entropy production. However, we have shown that the decrease is caused only by the negative perturbation applied to the sinking region, which destroys the initial circulation altogether (see Shimokawa and Ozawa, 2002, for details). In conclusion, the perturbations tend to shift the system to a state of higher entropy production, except when the perturbation destroys the initial circulation completely. This result supports the hypothesis that the nonlinear dynamic system tends to move to a state with higher entropy production (and eventually a MEP state) by producing active circulation in the system when triggered by perturbations.

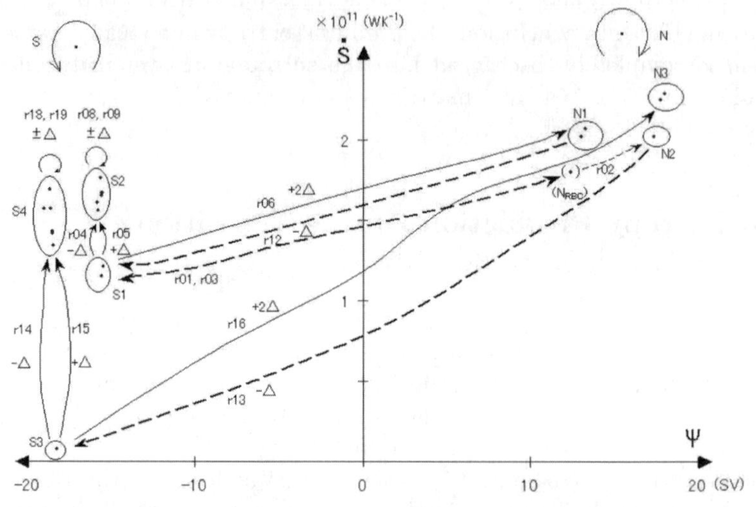

Fig. 10.4. The relationship between transitions among multiple steady states and rates of entropy production. The vertical axis ($\dot{S}$) indicates the rate of entropy production, and the horizontal axis (Ψ) shows the maximum value of the zonally integrated meridional stream function for the main circulation. The dots correspond to the steady states (initial and final states) of each experiment. The circles surrounding the dots show the circulation pattern (e.g., N1). The arrows show the direction of the transitions. The symbols beside the arrows show the experiment number and the perturbation used in the experiment (e.g., r04 and $-\Delta$)

10.6 Entropy Production During Evolution of Structure

Figure 10.5 shows the time evolution of the circulation during transition experiment r14; the upper panel is a time series of the rate of entropy production during the transition, and the lower panels show the structure of the circulation in terms of the zonally integrated meridional stream functions at certain years indicated with arrows in the upper panel. In r14, a negative salt perturbation ($-\Delta$) is applied to the Northern Hemisphere (46°–70° N, see marks in Fig. 10.5) for the first 500 years. The negative salt perturbation tends to intensify the southern sinking circulation. In fact, the initial southern circulation S3 develops into a more intense and deep circulation S4 (Figs. 10.2 and 10.4). The circulation has deepened after a rapid collapse of the weak bottom water circulation (B) at around year 690. This collapse is preceded by a collapse of a small circulation cell (A) beneath the main circulation at around year 640 (Figs. 10.5c, d). The main circulation develops through these successive changes, and the resultant rate of entropy production increases. One can also see that the circulation is stabilized with its evolution, and large oscillations in entropy production that exist in the initial circulation are considerably reduced during the transition. In conclusion, the rate of entropy production in the open ocean system is largely enhanced by the development of the circulatory structure in the system.

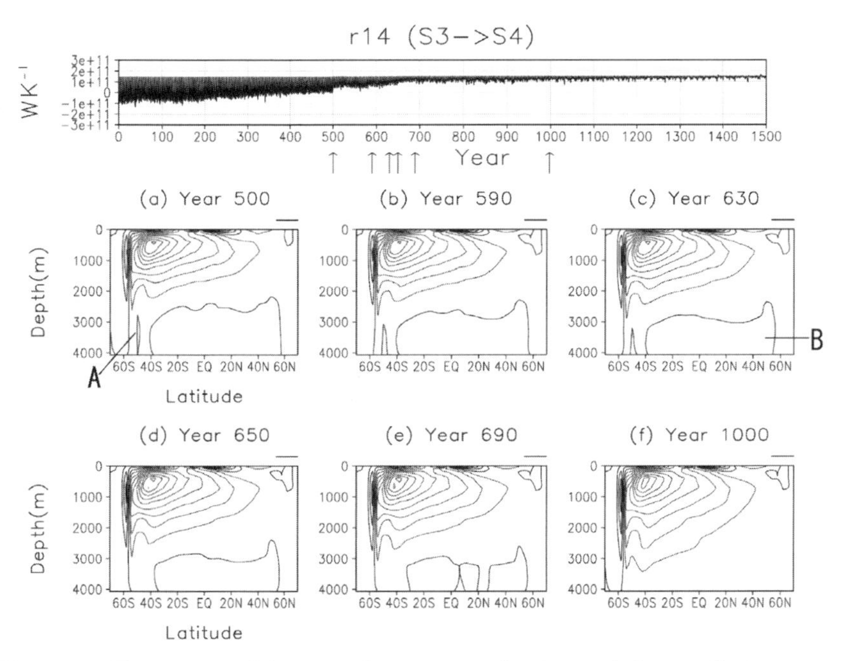

Fig. 10.5. Time series of the rate of entropy production and the zonally integrated meridional stream functions at year **a** 500, **b** 590, **c** 630, **d** 650, **e** 690, and **f** 1000 for r14. The unit of entropy production rate is W K^{-1}. The unit of meridional stream function is SV (10^6 m^3 s^{-1}). The contour interval of meridional stream function is 2.0 SV. The horizontal lines at the upper right side of each figure show the latitude range to which the perturbations are applied. In r14, the system moves from S3 to S4

10.7 Analogy Between Ocean System and Living System

Both the ocean system and living systems seem to share a common feature, that is, the development of a circulatory structure. With respect to living systems, Boltzmann (1886) once suggested that "the general struggle for existence of life is a struggle for entropy". His suggestion was later examined by Lotka (1922), Schrödinger (1944), Forrest and Walker (1964), Ulanowicz and Hannon (1987), Aoki (1991), and Sawada (1994). In this final section, we discuss the analogy between the ocean circulation and living organisms, and explore some basic thermodynamic properties of living organisms and their evolution.

We have seen in the preceding sections that time evolution of the circulatory structure in the ocean system is associated with an increase in the entropy production rate. By producing the highly organized circulatory structure, the system succeeded in producing a higher rate of entropy production (Fig. 10.6a). The same should hold for the development of a living organism. Through the development of a system from a fertilized egg to an adult, a well-organized circulation of body fluids (i.e., blood) emerges. By using this circulating blood as a working substance, the living system attains higher exchange rates of heat and materials (e.g., oxygen and food) with its surroundings, which should result in a higher rate of entropy production. The rate of entropy production is likely to be at its maximum in the living system's most mature state (Aoki, 1991, 1998).

There is, however, a difference between an ocean system and a living system. The circulation of the former can last perpetually, provided that a temperature difference exists in the surrounding system (i.e., the surrounding system is in a non-equilibrium state). In contrast, life activity and its entropy production decrease with old age, ending in death (Aoki, 1998). The existence of death seems to be an apparent contradiction to MEP, and it should be examined more carefully whether the conditions of MEP apply to the growth processes of individual organisms.

Let us consider the development of an internal structure of a species during a considerably long period of time, rather than that of an individual living system during that individual's short lifetime. The development of a structure in individuals of a species caused by differential reproduction processes on the geologic time scale is known as "evolution" after the significant work of Darwin (1859). According to Darwin's theory, all species present must have evolved from simpler species – ultimately from a unicellular organism in the primeval stage of life on Earth. At that stage all living systems might have been unicellular. The exchange rates of heat and materials with the outside environment are low for these systems, since only diffusion can take place (Fig. 10.6c, left). This diffusion-only state is unstable in the sense that the rate of entropy production is lower than that in the circulation state or in a state with dissipative structures (Fig. 10.6a, b). When unicellular systems started to aggregate with each other, and succeeded in producing a circula-

a Fluid circulation

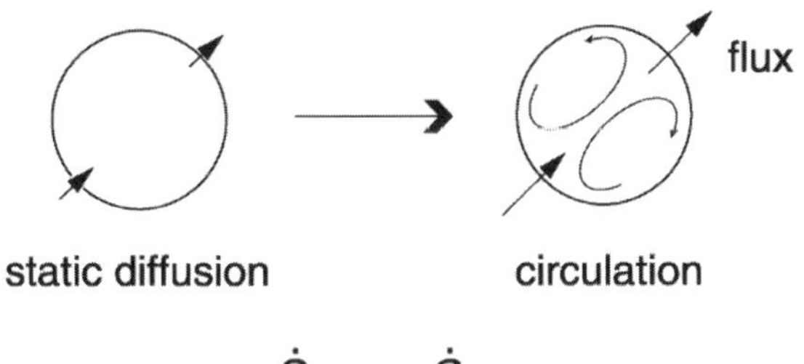

b Dissipative structure

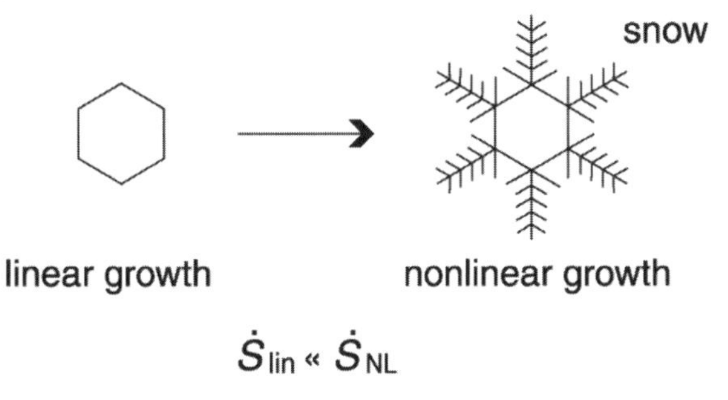

c Evolution of living systems

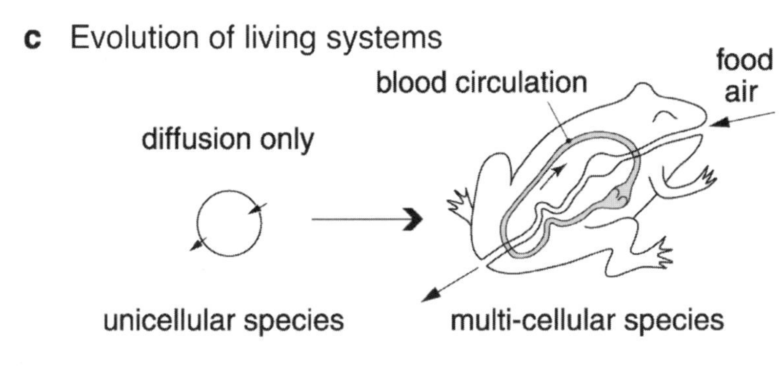

Fig. 10.6. An analogy among fluid circulation, dissipative structure, and the evolution of a living system. Each system produces a highly organized structure from a less organized primitive one in a consistent manner with the thermodynamic principle (see text for details)

tory structure of fluids by differential reproduction, this should have resulted in higher rates of entropy production (e.g., reflected in an increase in the basal metabolic rate with evolutionary younger species, Zotin 1984, also see connection to high concentrations of atmospheric oxygen, Catling, this volume). This development from unicellular to multi-cellular species is called the "Cambrian explosion", and occurred about 500 million years ago (Zhuravlev and Riding 2001; von Bloh et al. 2003; Catling, this volume). This surprising consequence of evolution is, however, consistent with the thermodynamic concept, since the rate of entropy production is much higher for multi-cellular species (Fig. 10.6c).

It should be noted here that the differential reproduction of living systems and the resultant evolution are made possible by the deaths of pre-existing living systems. Evolution would have been impossible if living systems had possessed eternal life. Even though the death of each individual living system briefly reduces entropy production locally, it also has a vital effect on evolution, allowing a species as a whole to attain a higher rate of entropy production in the longer time scale (Fig. 10.6c). In this sense, the death of each individual is inevitable and essential to the evolution of the whole species as well as to entropy production. And by differential evolution of many species on the Earth, the total rate of entropy production by all life seems to have grown to a considerable level by producing preferred distributions of species around the world. The entropy produced by living systems and by the ocean system is eventually discharged into space through emission of long-wave radiation (see, e.g., Ozawa et al., 2003). It is therefore possible to say that the evolution of all living systems and the planetary circulation are controlled by the universal requirement of entropy production in the Universe.

References

Aoki I (1991) Entropy principle for human development, growth and aging. J Theor Biol 150: 215–223.

Aoki I (1998) Entropy and exergy in the development of living systems: A case study of lake-ecosystems. J Phys Soc Japan 67: 2132–2139.

Boltzmann L (1886) Der zweite Hauptsatz der mechanischen Wärmetheorie. Gerold, Vienna (Republished in Populäre Schriften, JA Barth, Leipzig, 1905; Vieweg, Braunschweig-Wiesbaden, 1979).

Bryan F (1986) High-latitude salinity effects and interhemispheric thermohaline circulation. Nature 323: 301–304.

Darwin C (1859) The origin of species by means of natural selection. J. Murray, London.

Dewar R (2003) Information theory explanation of the fluctuation theorem, maximum entropy production and self-organized criticality in non-equilibrium stationary states. J Phys A 36: 631–641.

Forrest WW, Walker DJ (1964) Change in entropy during bacterial metabolism. Nature 201: 49–52.

Glansdorff P and Prigogine I (1964) On a general evolution criterion in macroscopic physics. Physica 30: 351–374.

Grassl H (1981) The climate at maximum entropy production by meridional atmospheric and oceanic heat fluxes. Q J Roy Meteorol Soc 107: 153–166.

Gregg MC (1984) Entropy generation in the ocean by small-scale mixing. J Phys Oceanogr 14: 688–711.

Hill A (1990) Entropy production as the selection rule between different growth morphologies. Nature 348: 426–428.

Lorenz RD, Lunine JI, Withers PG, McKay CP (2001) Titan, Mars and Earth: entropy production by latitudinal heat transport. Geophys Res Lett 28: 415–418.

Lotka AJ (1922) Contribution to the energetics of evolution. Proc Nat Acad Sci 8: 147–150.

Marotzke J, Willebrand J (1991) Multiple euilibria of the global thermohaline circulation. J Phys Oceanogr 21: 1372–1385.

Martyushev LM, Seleznev VD, Kuznetsov IE (2000) Application of the principle of maximum entropy production to the analysis of the morphological stability of a growing crystal. J Exper Theor Phys 91: 132–143.

Noda A, Tokioka T (1983) Climates at minima of the entropy exchange rate. J Meteorol Soc Japan 61: 894–908.

Ozawa H, Ohmura A (1997) Thermodynamics of a global-mean state of the atmosphere – a state of maximum entropy increase. J Climate 10: 441–445.

Ozawa H, Ohmura A, Lorenz RD, Pujol T (2003) The second law of thermodynamics and the global climate system – A review of the maximum entropy production principle, Rev Geophys 41: 1018. doi: 10.1029/2002RG0000113.

Pacanowski RC (1995) MOM2 documantation, user's guide and reference manual. GFDL Ocean Group Technical Report 3, Princeton, NJ, USA.

Paltridge GW (1975) Global dynamics and climate – a system of minimum entropy exchange. Q J Roy Meteorol Soc 101: 475–484.

Paltridge GW (1978) The steady-state format of global climate. Q J Roy Meteorol Soc 104: 927–945.

Prigogine I (1955) Introduction to thermodynamics of irreversible processes. John Wiley & Sons Inc., New York.

Sawada Y (1981) A thermodynamic variational principle in nonlinear non-equilibirium phenomena. Prog Theor Phys 66: 68–76.

Sawada Y (1994) A scaling theory of living state. Physica A204: 543–554.

Schrödinger E (1944) *What is Life?* Cambridge University Press, Cambridge.

Shimokawa S, Ozawa H (2001) On the thermodynamics of the oceanic general circulation: entropy increase rate of an open dissipative system and its surroundings. Tellus A53: 266–277.

Shimokawa S, Ozawa H (2002) On the thermodynamics of the oceanic general circulation: irreversible transition to a state with higher rate of entropy production. Q J Roy Meteorol Soc 128: 2115–2128.

Stommel H (1961) Thermohaline convection with two stable regimes of flow. Tellus 13: 224–230.

Suzuki M, Sawada Y (1983) Relative stabilities of metastable states of convecting charged-fluid systems by computer simulation. Phys Rev A27: 478–489.

Ulanowicz RE, Hannon BM (1987) Life and the production of entropy. Proc R Soc Lond B 232: 181–192.

Vanyo JP, Paltridge GW (1981) A model for energy dissipation at the mantle-core boundary. Geophys J R Astr Soc 66: 677–690.

Von Bloh W, Bounama C, Franck S (2003) Cambrian explosion triggered by geoshere-biosphere feedbacks. Geophys Res Lett 30: 1963. doi: 10.1029/2003GL017928.

Woo H-J (2002) Variational formulation of nonequilibrium thermodynamics for hydrodynamic pattern formations. Phys Rev E66: 066104-1-066104-5

Zhuravlev AY, Riding R (eds.) (2001) The Ecology of the Cambrian radiation, Columbia University Press, New York.

Zotin AI (1984) Bioenergetic Trends of Evolutionary Progress of Organisms. in: Lamprecht I, Zotin AI (eds) Thermodynamics and Regulation of Biological Processes, Walter de Gruyter, Berlin, Germany, pp 451–458.

11 Entropy and the Shaping of the Landscape by Water

Hideaki Miyamoto[1], Victor R. Baker[2], and Ralph D. Lorenz[3]

[1] Department of Geosystem Engineering, University of Tokyo, Tokyo, Japan
[2] Department of Hydrology and Water Resources, University of Arizona, Tucson, AZ, USA
[3] Lunar and Planetary Laboratory, University of Arizona, Tucson, AZ, USA

Summary. We explore applications of thermodynamics to hydrology, in particular the application of extremization principles to self-organized river networks. Two thermodynamic principles have been applied to river networks: (1) the most probable state of a system is that where its configurational entropy is a maximum, corresponding to dissipation spread evenly throughout the network, and (2) the principle of minimum total energy dissipation, similar to the principle of minimum entropy production. We also discuss the power-law characteristics that are observed in river networks and show how they arise in model networks. We also note the application of these principles to shoreline profiles.

11.1 Introduction

Thermodynamics and Hydrology have always been connected: Carnot (1824) likened the downhill flow of water and the flow down a temperature gradient of heat. More recently, further analogies have been developed, and certain extremal properties of river networks have been recognized that bear discussion in the framework of thermodynamic extremizations described elsewhere in this book. Furthermore, a probabilistic entropy can be defined for a drainage network and is a characteristic parameter of the network's appearance, which exhibits a high degree of regularity and spatial organization.

Of course, the very fact that the Earth has a hydrological cycle at all attests to disequilibrium in the hydrosphere. Were the delivery of net radiative energy to evaporate water to stop, and rain to cease falling, rivers and lakes would eventually drain, and the distribution of water would converge to a minimum energy state with respect to potential energy. On the other hand, net energy input leads to a hydrological cycle where part of the fluid inventory is always in motion by virtue of evaporation and precipitation, requiring work and thus resulting in entropy production (see e.g., Pauluis, this volume).

In a fluvial system, the set of physical factors includes many variables such as the amount of water and sediment to be carried, the fluid friction, and the fluvial transport capacity. Hydrologists have tried to find equations to determine these factors and drew the conclusion that several degrees of freedom always remain: A fluvial system can adjust its dimension and dynamics to a given slope in several ways. Its complex but highly regular appearance, however, inspired researchers to seek for fundamental principles that explain the morphological characteristics. The use of an extremal hypothesis, first advocated by Leopold and Langbein (1962), provides an additional relationship to constrain the river channel morphologies.

Molnár and Ramírez (1998) suggested that two optimality hypotheses might be reasonably addressed for the morphology of the river network: the local channel properties which will be adjusted toward an optimal state, and the global property that a river network adjusts its topological structure toward a state in which the total rate of energy dissipation in the network is a minimum. The local optimality hypothesis addresses mainly the relatively short-term adjustment of internal channel geometry and sediment load, and was pioneered by Leopold and Langbein (1962), which led a wide range of applications such as Chiu (1986), Yang and Song (1986), Molnár and Ramírez (1998). The global optimality hypothesis addresses the long-term adjustment of the topological structure of the river network in response to geologic driving forces, continuous erosion, and long-term changes in the runoff amount and sediment supply (Rodríguez-Iturbe et al. 1992; Rinaldo et al. 1992). It can be seen as being related to the Minimum Entropy Production (MinEP) principle as discussed in the introduction of this book, with a fixed flux of precipitation entering the river network and a fixed flux of discharge exiting the network.

In the following we first give a brief historical account of the application of thermodynamic concepts to hydrology (Sect. 11.2), then describe scaling laws in hydrology (Sect. 11.3) and how they can be understood from fractal treatment of optimal channel networks (Sect. 11.4). Further applications of these ideas to shoreline profiles are briefly discussed in Sect. 11.5 and we close with a conclusion of the potential applications of thermodynamics in hydrology.

11.2 Early Work by Leopold and Langbein

Carnot's analogy was drawn in reverse by two hydrologists, Luna Leopold and Walter Langbein, in a 20-page report 'The Concept of Entropy in Landscape Evolution' published by the US Geological Survey in 1962. Among the ideas in this paper are that:

1. the height in a river network (or hydraulic head) is analogous to temperature, and that analogous properties to heat flow and entropy can be defined;
2. likely states of a system are those with a maximum entropy (they make this observation without apparent reference to Jaynes' work, identifying maximum entropy configurations of a system as 'most probable' only 5 years earlier – see the chapter in this volume by Dewar);
3. certain power-law properties of a drainage network arise;
4. a random-walk model of river networks yields comparable properties with observed ones, and thus river network formation and evolution may be considered a statistical problem.

They drew an analogy between on the one hand a landscape with water added at different elevations and draining stepwise through a river network, and a cascade of steam engines whose work is dissipated by brakes, with heat added to each engine, and the waste heat from each feeding the engines lower down in the set (Fig. 11.1)

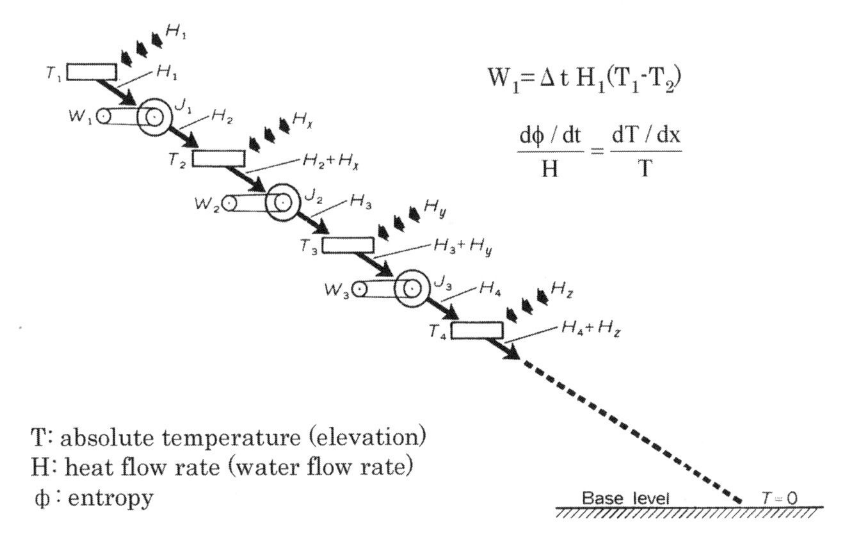

Fig. 11.1. Thermodynamic engine model of a steady state river system. A river system may be considered in dynamic equilibrium, though it is an open system in which energy is being added in some places while in other places energy is being degraded to heat. In this case, the rate of outflow of entropy, represented by the dissipation of energy as heat, equals the rate of generation of entropy which is represented by the energy gradient toward base level. The above model consists of a series of perfect engines (or a river reach), J_1, J_2, etc., operating between heat sources and sinks designated by their absolute temperatures (or elevations), T_1, T_2, etc. The heat sources (the schematic in Fig. 11.1 has only one sink, since this is the typical case for a river network, but the idea is easily generalized) H_1, H_2 etc. correspond to mass flows of fluid delivered to the system, e.g., by precipitation

They pointed out the widespread use of a 'principle of least work' in structural engineering to solve "statically indeterminate" problems in the stress analysis of frame structures. Society expects builders of bridges to be careful and conservative, and yet the discipline accepts an extremization principle to resolve an otherwise unsolvable problem, without any fundamental logic other than that it works. It is true that systems in general tend to minimize their free energy in steady-state, but to consider that a pin-jointed frame like the latticework of a bridge will attain such a state somehow tacitly acknowledges that it is a dynamic system that can have many states, and evolves between them (albeit only with some shaking). This seems an alien and uncomfortable notion for the otherwise rigid discipline of structural engineering, perhaps a reason why it is taught and used with little comment.

A corollary of the maximum entropy idea is that, given a network with a certain number of links, and a certain dissipation that must be distributed across the network (however that global dissipation happens to be defined), the maximum entropy state is one that has the dissipation spread uniformly across the network. In other words, each link has the same dissipation as every other one. There are more possible permutations that distribute the 'quanta' of dissipation evenly or near-evenly than there are which concentrate them.

They use such an entropy maximization argument to show that the most probable profile of a river bed is exponential and similarly derive power-law scaling relationships for a number of river properties such as speed-discharge, depth-discharge and so on. They show that the resultant scaling relationships are quite comparable with observations, many of these made by Leopold himself.

Leopold and Langbein provide a demonstration that random-walk (i.e., Brownian motion) models of rivers, with streams starting from arbitrary points, coalesce to form networks with comparable stream-length to stream-order relationships as those observed in nature. (Their simulation is remarkable in that it was performed without the aid of a computer!)

They also observe that only a few trials are needed for the system to attain properties characteristic of those in nature and those with a maximum entropy, and that a very large number of trials are needed to distinguish those at the very peak from those merely close to it.

Self-organization in river networks is evident in the dendritic patterns that are so familiar to us and thermodynamics offers a framework in which this self-organization can be understood as an attempt by the rivers to control the dissipation of the potential energy they are supplied. Their paper closes with the remark:

> *"Whether or not the particular inferences stated in the present paper are sustained, we believe that the concept of entropy and the most probable state provides a basic mathematical conception which does deal with relations of time and space."*

11.3 Scaling Laws in Hydrology

As Leopold and Langbein (1962) theoretically explained the background, the power-law scaling relationships exist in a number of river properties. One of the most commonly cited is the so-called Hack's law. Hack (1957) demonstrated the applicability of a power function relating length L and area A for streams, and proposed such as $L \sim A^{\alpha}(\alpha = 0.6)$ and $Q \sim A^{\beta}(\beta = 1.0)$ for streams in Virginia, where Q is the flow discharge. Stable channels are also known to be reasonably described by "regime theory": the averaged width $w = a\,Q^{1/2}$, the averaged depth $d = c\,Q^{1/3}$, and slope $s = t\,Q^{-1/6}$, where a, c, and t are parameters related to sediment composition, vegetation, channel roughness, and so on. Yang (1971) also made an analogy between thermodynamic and fluvial systems, with elevation analogous to temperature, and suggested that minimum energy dissipation rate is a general theory in fluvial hydraulics (e.g., Yang and Song 1986).

The slope scaling of the power law cannot extend indefinitely, as that would imply infinite slopes at small scales. A break in scaling at scales smaller than about 0.6 km was observed by Mark and Aronson (1984). This scale may be interpreted as one in which domination of sediment transport changes from fluvial processes to non-fluvial processes, and the precise value of scales will vary for different landscapes and geological conditions.

In addition to describing how channels adjust their shapes toward a steady state in which stream power or energy dissipation is at a minimum, the theory of optimal energy expenditure has been used to describe larger scale structure of optimal river networks. A set of principles that govern the evolution of river networks to the optimal state (optimal channel networks, or OCNs) is proposed to describe the formation of optimal topological structures by minimizing the total rate of energy expenditure (Rodríguez-Iturbe et al. 1992). Before proceeding to further discussions, we here review the efficiency, or the energy expenditure, of a river network.

The energy expenditure of a fluvial network can roughly be estimated as follows. The energy expenditure per unit time, P, in any link can be written as $P \sim Q\Delta z$, where Δz is the topographic difference in a link and is equal to $\nabla(zL)$. Put simply, the dissipation per unit length is the product of mass flowrate and slope. Gravitational acceleration is assumed constant and is omitted from these expressions. Tarboton et al. (1989) suggested that the group sample variances, Var $[\nabla z(A)]$ show power-law scaling proportional to $A^{-\theta}$, where θ is in the range $0.53 - 0.78$. If we assume $\nabla z(A) \sim A^{0.5}$ and $Q \sim A$ (uniform unit rainfall occurs at any site at any time,) we obtain $P \sim Q^{0.5}L$.

The above equation is derived from a somewhat stochastic method, but can also be obtained by hydraulic models. The overall gravitational driving force F for the water flow on a shallow slope in steady state can be written as $F \sim \rho g\, d\, L\, w \sin\theta$, where ρ is the density, g is the gravity, and θ is the slope angle ($\sin\theta \sim \nabla z$, where ∇z is the slope). The resisting force F_r may be mainly concentrated at the boundary of the flow, so if we assume a constant shear stress τ throughout the cross section, we can write $F_r \sim \tau(2d + w)\,L$. From a steady state flow assumption ($F = F_r$), τ can be written as $\tau = \rho g R \nabla z$, where R is the hydraulic radius defined by the cross sectional area A_w and the wetted perimeter P_w as $R = A_w/P_w$.

It is generally accepted that a fluvial system dissipates its energy by (1) friction at the boundary, (2) internal momentum dissipation due to viscosity, turbulence, and particle collisions, (3) eroding the channel bed, and (4) transporting sediments. Although these four factors have been studied by many researchers, it is difficult to quantify the rate of energy dissipation in general expressions. Therefore, the following discussion depends on empirical equations which are, however, qualitatively explained by physics of turbulent fluid dynamics.

Because the Reynolds number of river water is high enough, the river water would be reasonably considered as fully turbulent. In this case the velocity v at a certain depth can be written as $v = v_a + v^+$, where v_a is the time-averaged velocity and v^+ is a fluctuation in velocity. The vertically carried momentum per unit time and unit area can be considered to the turbulent shear stress, which is written as $\tau = -\rho(v_a v^+ + v^+ v^+)$, with the assumption of the isotropic fluctuation. With the mixing length theory by Prandtl, a time-averaged value for the turbulent shear stress is modeled as $\tau \sim v^2$. On this basis, empirical equations used in the engineering field are theoretically validated and is typically expressed as $\tau \sim C_f \rho v^2$, where C_f is a dimensionless resistance coefficient related to factors (1) and (2) listed above.

The two equations related to τ from above, the energy loss due to the friction per unit weight of flow per unit length can now be written as $C_f v^2/(Rg)$. Another expenditure of energy is related to the removal and transportation of sediments. However, this is a complex function of soil and flow properties and may not be expressed by simple equations. One approach is based on the assumption that this function is proportional to $K\tau^m$, where K is a parameter depending on soil and fluid properties and m is a constant (factors (3) and (4) from above are considered in combination). In this case, the overall energy expenditure in a channel with flow rate Q becomes

$$P = C_f \rho \frac{v^2}{R} QL + K\tau^m P_w L = \frac{QL}{d} P' + LdP'' \tag{11.1}$$

where $P' = C_f \rho v^2 + K C_f^m \rho^m v^{2m-1}$ and $P'' = 2C_f \rho v^3 + 2K C_f^m \rho^m v^{2m}$. This equation has the following important implications as discussed in Rodríguez-Iturbe et al. (1992):

1. The two factors for energy expenditure per unit length (an empirical frictional expenditure and a removal and a transportation of the sediment) are complicated functions of water velocity in different exponential parameters. Therefore, if we assume constant energy expenditure throughout the channel, it requires a constant velocity anywhere in the network.
2. The above equation achieves a minimum with a zero derivative of P with respect to depth d. This yields $Q \sim d^2$. In this case, the optimal power expenditure at any link can be written as $P = k\,Q^{0.5} L$, which is also suggested above from stochastic equations.

Figure 11.2 shows the difference in total rate of energy expenditure of different topologic patterns, illustrating that a tree-like pattern reduces the total rate of energy expenditure. This implies that a tree-like path more likely satisfies the above two criteria, justifying the above assumptions and approach to explain topologic characteristics of natural river systems.

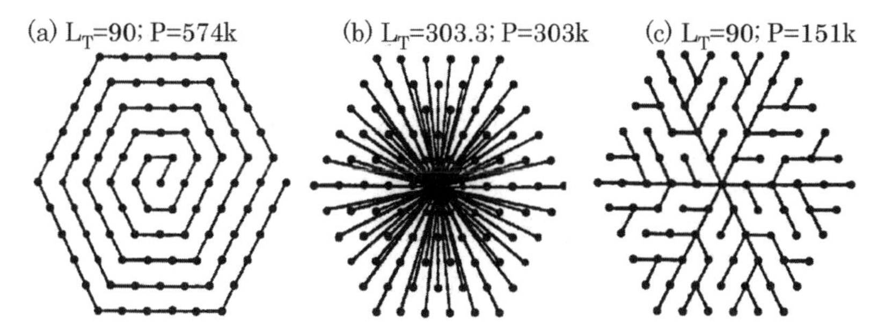

Fig. 11.2. Different patterns of connectivity for a set of equally spaced points connected to a common outlet. L_T is the total path length and P is the total rate of energy expenditure with input flow at any node being equal to 1 (from Rodríguez-Iturbe et al. 1992). Note that a treelike pattern yields a much smaller total rate of energy expenditure

11.4 Thermodynamics of Fractal Networks

Leopold and Langbein (1962) first carried out modeling studies of drainage basins by simulating the development of drainage networks through random walks in a rectangular region. In their approach, every square tile is drained and the flow path has an equal change of flowing to any of the adjacent tiles with a restriction of no flow into the reverse direction.

Shreve (1966) introduced the concept of topologically random population networks to geomorphology. In this model, all topologically distinct channel networks with a given number of sources are equally likely. As we will see below, this is the limit case of the model of optimal channel networks with a Boltzmann distribution for $T \to \infty$.

Random-walk and random-topology models are basically statistical models to explain static morphological characteristics and produce fractal structures closely resembling many characteristics of real river networks. In addition to describing channels in equilibrium, the theory of optimal energy expenditure has been used to describe the evolution of river networks to the optimal state. Rodríguez-Iturbe et al. (1992) introduced the concept of least energy expenditure toward a pre-established optimal state leading to the modeling of optimal channel networks (OCNs). OCNs are quite intriguing not only because they exhibit remarkable similarities with river networks extracted from digital elevation models (DEMs) but also because their scaling properties of the energy and entropy are supported by thermodynamic rationale as discussed by Rinaldo et al. (1996). Here we will briefly review their seminal works.

Optimal channel network (OCN) configurations are obtained by minimizing the total rate of energy expenditure in the river system as a whole and in its parts. The hypothesis that the process of network formation, as well as of a broader class of patterns in nature, is characterized by minimum total energy dissipation of the resulting three-dimensional geometry and topology.

Following Rodríguez-Iturbe et al. (1992), three principles are postulated:

1. Equal energy expenditure per unit area of channel: The channel achieves an optimal condition anywhere regardless of its topological structure.
2. Minimum energy expenditure in any link: The energy expenditure for any link of the network should be a minimum.
3. Minimum energy expenditure in the network: This is achieved by a traveling salesman-like algorithm (i.e, the 'simulated annealing' approach, or Metropolis algorithm.) The algorithm randomly selects a site and perturbs its configuration to see the change in the energy expenditure. The probability of acceptance depends (Arrhenius-like) on the energy change incurred – if it reduces the energy of the system, the perturbation is more likely to be accepted. This procedure stops after a prefixed number of changes are rejected.

The thermodynamic basis behind the scaling properties of the energy and entropy of OCNs is described by Rinaldo et al. (1996). The OCN is obtained by selecting the spanning tree, characterized by a variable s that minimizes the Hamiltonian of the system defined as:

$$H_\gamma(s) = \sum_{i=1}^{L^2} A_i^\gamma \tag{11.2}$$

where i spans the L^2 sites occupied by a $L \times L$ square lattice, γ is usually considered as a constant parameter (~ 0.5 is often used from the physics of the erosional process), and A_i is the total contributing area at the ith site of the lattice. For each spanning tree configuration, s, a Boltzmann-like probability can be defined as:

$$P(s) \propto \exp(-H_\gamma(s)/T) \tag{11.3}$$

where T is a parameter resembling Gibbs' temperature of thermodynamic systems.

Here we assume S is the set of spanning loopless trees of drainage channels rooted in a given point, E is a given energy level, and $N(E)$ is the number of different spanning trees s for which $H_\gamma(s) = E$. Defining the thermodynamic entropy as $\sigma(E) = \log N(E)$, we can write

$$P(H_\gamma(S) = E) = \sum_{s:H_\gamma(S)=E} P(s) \propto \exp(-F(E)/T) \tag{11.4}$$

where $F(E)$ is a free energy written as $F(E) = E - T\_(E)$. $F(E)$ can be written for OCNs with $\gamma \geq 0.5$ as $F(E) \propto L^{2+\delta} - aTL^2$, where a is a constant (see Rinaldo et al. (1996) for details). Therefore, when the configuration s minimizes H_γ, it also minimizes $F(E)$, regardless the value of Gibbs' parameter if the system is large enough. In other words, the most probable spanning tree configurations determined by minimizing the free energy can be equally well obtained by minimizing the energy, provided that L is large enough. Real fluvial networks usually show their drainage patterns without geologic controls over domains, though it is not true if we see channels at relatively smaller scales. This may be explained by the fact that the natural networks are in conditions that well approximate the thermodynamic limit ($L \to \infty$), and this gives the same conditions for yielding the minimum free energy with zero-temperature approximation ($T \to \infty$) as OCNs which explains many observational characteristics of networks regardless of geology, vegetation, or other factors.

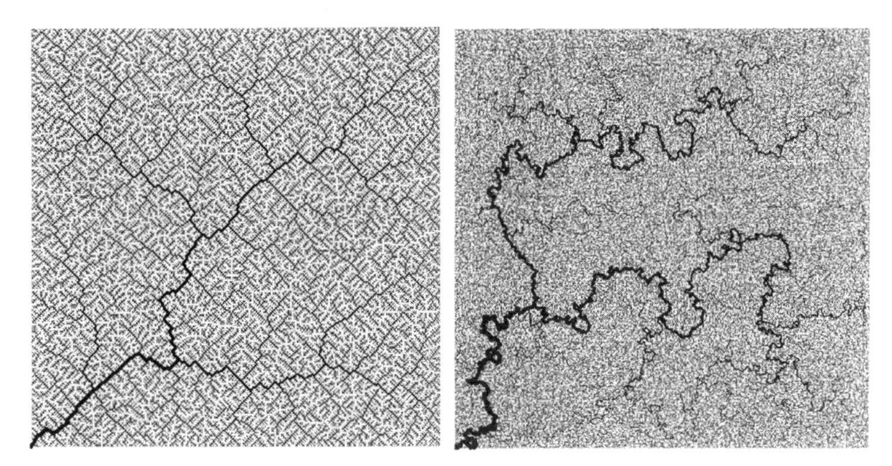

Fig. 11.3. A configuration of a "cold" 256×256 OCN (*left*) and "hot" OCN (*right*) (after Rinaldo et al. 1996)

If $T > L^{\delta}/a$, i.e., a 'hot' network, then entropy dominates over energy. The example calculation of this case made by Rinaldo et al. (1996) shows fractal characteristic (Fig. 11.3) but no fractal properties match those values of production zones of natural channels, though a certain morphological similarity exists. At lower temperature, the appearance of the network after a given number of iterations more closely resemble those of real basins. However, similar to the thermodynamic interpretation discussed above, the hot condition might correspond to the case for which the characteristic size L is relatively small, and, therefore, geologically constrained. While, as Rodríguez-Iturbe and Rinaldo (1997) suggest, a clear geomorphologic assessment of the physical meaning of the mutational parameter T is lacking, we can nevertheless conjecture that hot patterns may represent patterns occurring in low-relief areas. The low relief means that there is a small 'activation energy' for changes to the network, and thus abrupt changes in direction such as meanders can spontaneously occur.

There is much that remains to be explored. Investigating the effects of Gibbs' and gamma parameters and iterations on the network characteristics such as their 'maturity', fractal dimensions, and other statistical properties (tortuosity, for example), will be useful when we consider a more general formation of topographic patterns curved for an unknown period by unknown agents (Fig. 11.4). On planetary surfaces, there are many network features which can be explained by multiple – i.e., contradictory – hypotheses with different physics (e.g., is a given valley carved by water flow, glacial flow, lava flow or underground sapping?). A pertinent example is the planet Mars, where the debate continues as to whether certain features such as crater-wall gullies and valleys were formed by water or by debris flows mobilized by sublimating carbon dioxide deposits. Geomorphology tends to have a somewhat subjective character, in that different workers may assign different formation mechanisms or even labels to a given feature. Quantitative metrics, such as the network parameters we have discussed here, offer an objective criterion by which features can be classified.

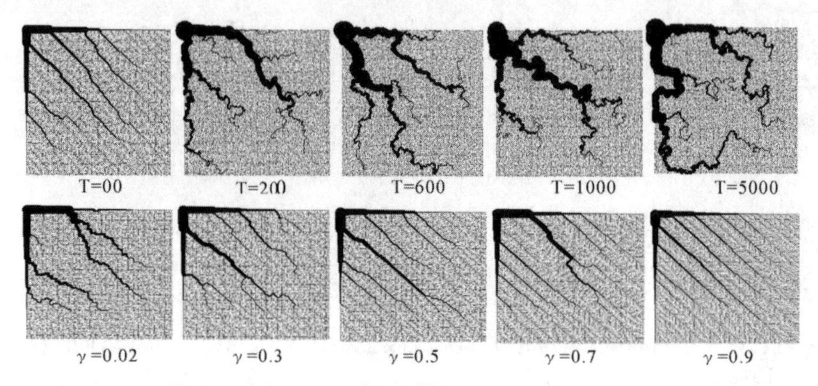

Fig. 11.4. Networks obtained for different T (*top*) and γ (*bottom*) parameters. All networks are from the same initial conditions and after 10^6 iterations with 100×100 cells. γ is fixed at 0.5 for the *top figures*, and T is fixed at 0.0 for the *bottom figures*

Furthermore, an extension of this work would be to relate the observed network characteristics with specific formation parameters: a given network may have a volume that implies a certain flowrate-time product or at least a finite set of such products. But the thermodynamic character of the network may yield quantitative insights as to whether the network was carved slowly by continuous yet weak processes, or in a small number of short but violent events.

11.5 Entropy and Shoreline Profiles

Various workers have also explored the application of very similar ideas to shoreline profiles (see Cowell et al. 1999 for a summary). Specifically, the shoreface forms by a set of processes which cause sediments to diffuse to eliminate work gradients. This equal-dissipation argument as before leads to a maximum-entropy concept. One should note the distinction (and thus the corresponding ambiguity in which should apply or be dominant) between dissipation per unit area of seabed, and dissipation per unit volume. The latter may be more appropriate in the turbulent surf zone, while in deeper water the dissipation is concentrated at the seabed and thus dissipation per unit area is more appropriate. An equilibrium profile of depth $h(x)$ of the form $h = A\,x^m$ results, with A being a constant that relates to sand grain size, and m is an exponent with a value that might be expected to be $\sim 2/3$ assuming constant dissipation per unit volume, or $\sim 2/5$ for constant dissipation per unit area. Field observations indicate $2/5 < m < 4/5$, with the lower values characteristic of wave-reflecting beaches, and the higher values corresponding to beaches with dissipative surf zones.

11.6 Concluding Remarks

We have given an overview of some energetic and thermodynamic principles that have been applied to hydrology. We note there are currently two distinct analogies, though we suspect they should ultimately be related. The first principle states that a fluvial system may be considered as an engine driven by the supply of water at high elevation flowing to low elevation. In this sense elevation, or hydraulic head, is a direct analogue of temperature for a heat engine. Secondly, the configuration of a river network can be described by certain statistical properties, notably an entropy. The notion of minimum energy expenditure in the network should then correspond to the principle of minimum entropy production, where the prescribed boundary conditions do not allow for much flexibility (see e.g., Kleidon and Lorenz, this volume). However, since much of the dissipation of energy in river networks is related to turbulence, Maximum Entropy Production should also be applicable, although the detailed application is not clear and needs further investigations.

If the thermodynamic background to the hydrological shaping of the landscape becomes sufficiently understood, practical applications may be developed. In particular, statistical properties of hydrologic networks may provide quantitative means of classifying networks and thereby understanding the geomorphological processes by which they formed. The relationship of the observed 'maturity' of a river system with the Gibbs' parameter (i.e., network temperature) may be a fruitful avenue of enquiry.

Finally, we note that exploring the thermodynamic concepts of fluvial geomorphology also appear useful for shoreline processes. Their generality may make these ideas quite fruitful for investigating networks on other planetary bodies, where the specific mechanisms and working substances may be different from Earth, but the aggregate effects are similar.

References

Carnot S (1824) Réflexion sur la puissance motrice du feu et sur les machines propres á développer cette puissance, Bachelier, Paris, 118pp. (English translation: reflexions on the motive power of fire, Manchester Univ. Press, Manchester, UK, 230pp).

Chiu CL (1986) Entropy and probability concepts in hydraulics. J Hydraulic Eng 113: 583–600.

Cowell PJ, Hanslow DJ, Meleo JF (1999) The Shoreface. in Short AD (ed) Handbook of Beach and Shoreface Morphodynamics. Wiley, Chichester, pp. 39–71.

Hack JT (1957) Studies of longitudinal profiles in Virginia and Maryland. US Geol Surv Prof Paper 294-B, US Government Printing Office, Washington DC.

Langbein WB, Leopold LB (1964) Quasi-Equilibrium States in Channel Morphology. Am J Sci 262: 782–794

Leopold LB, Langbein WB (1962) The Concept of Entropy in Landscape Evolution. US Geol Surv Prof Paper 500-A, US Government Printing Office, Washington DC.

Mark DM, Aronson PB (1984). Scale-dependent fractal dimensions of topographic surfaces: An empirical investigation with applications in geomorphology and computer mapping. Math Geol 16: 671–683.

Molnár P, Ramírez JA (1998) Energy dissipation theories and optimal channel characteristics of river networks. Water Resour Res 34: 1809–1818.

Rinaldo A, Rodríguez-Iturbe I, Rigon R, Bras RL, Ijjasz-Vasquez E, Marani A (1992). Minimum energy and fractal structures of drainage networks. Water Resour Res 28: 2183–2195.

Rinaldo A, Maritan A, Colaiori F, Flammini A, Rigon R, Rodríguez-Iturbe I, Banavar JR (1996) Thermodynamics of Fractal Networks. Phys Rev Lett 76: 3364–3367.

Rodríguez-Iturbe I, Rinaldo A, Ringon R, Bras RL, Marani A, Ijjasz-Vasquez E (1992) Energy dissipation, runoff production, and the three-dimensional structure of river basins. Water Resour Res 28: 1095–1103.

Rodríguez-Iturbe I, Rinaldo A (1997) Fractal River Basins: Chance and Self-Organization. Cambridge University Press, Cambridge, UK.

Shreve RL (1966) Statistical law of stream numbers. J Geol 74: 17–37.

Tarbonton DG, Bras RL, Rodoriguez-Iturbe I (1989) Scaling and elevation in river networks. Water Resour Res 25: 2037–2051.

Yang CT (1971) Potential energy and stream morphology. Water Resour Res 7: 311–322.

Yang CT, Song CCS (1986) Theory of minimum energy and energy dissipation rate. in Cheremisinoff ND (ed) Encyclopaedia of Fluid Mechanics, Gulf, Houston, TX, pp. 353–399.

12 Entropy Production in the Planetary Context

Ralph D. Lorenz

Lunar and Planetary Laboratory, University of Arizona, Tucson, AZ 85721, USA

Summary. In this paper I review some applications of nonequilibrium thermodynamics to planetary science. Of particular importance are the horizontal and vertical transports of heat in planetary atmospheres. It has been noted that Titan and Mars, like the Earth, appear to have equator-to-pole heat transports consistent with a Maximum Entropy Production principle. The transport of heat by convection in the atmospheres and interiors of the planets can be viewed in heat engine terms, and useful insights gained by considering the irreversibilities in these systems. Even bodies in space without an atmosphere can act as heat engines, their orbits being modified by the Yarkovsky effect wherein sunlight is downconverted into thermal radiation which is reradiated anisotropically – a rocket using thermal photons as propellant. Finally, spacefaring civilizations may seek to maximize the production of entropy from their parent stars by erecting a Dyson sphere in order to reject its power at a minimum temperature.

12.1 Equator-Pole Temperature Gradients of Planetary Atmospheres

Planets are objects that intercept low-entropy energy as sunlight (short wavelength and unidirectional) and reradiate it in a higher-entropy form (as longer wavelength heat, and in many directions, e.g., Aoki 1983). Because planets are round, high latitude regions receive less sunlight than the tropics except for planets with high obliquities, when the opposite is the case. On an airless body, there is no heat transport and there is therefore a substantial equator-to-pole temperature gradient and the planet is everywhere in radiative equilibrium. Planets with significant atmospheres (and oceans) permit significant heat flows which follow, and therefore mitigate, the temperature gradient. Were this heat flow performed with a near-infinite conductivity (or diffusivity) the planet would be isothermal.

As noted by Paltridge(1975) and in other chapters in this volume, the Earth's zonal climate can be reproduced remarkably well by assuming that these heat flows maximize the production of entropy. Simply put, if the heat flow F from tropical regions at temperature T_0 to cooler polar regions at T_1, the quantity $dS/dt = F/T_1 - F/T_0$ is maximized. Lorenz (1960) had noted before a related and essentially equivalent observation, that the Earth's climate maximizes the generation rate of Available Potential Energy (APE).

Many papers discussing the application of MEP note that the generality of the principle should allow it to apply to other planetary bodies, and indeed since the coincidence of the observed terrestrial climate state with one of MEP might be just that, coincidence, an evaluation of the principle on other planets would be an independent test of the applicability of MEP. For this purpose, Lorenz et al. (2001) presented a simple two-box model that captures the essence of climate on a round planet (similar to the one used in the introductory chapter, Kleidon and Lorenz, this volume). A low and high latitude zone each receive different amounts of sunlight, each reject heat as a function of their temperature, and heat is transported between the two boxes depending on an effective heat diffusivity D. By varying D, the effect of heat transport on entropy production and the equator-pole temperature gradient of different planetary atmospheres can be investigated.

12.1.1 Earth

For the present-day Earth, the insolation I_0 in the low-latitude box is about $I_0 = 240\,W\,\mathrm{m}^{-2}$, taking into account the albedo, while at higher latitudes we have $I_1 = 140\,W\,\mathrm{m}^{-2}$. Heat leaves the planet as thermal radiation from the two zones, with the outgoing emission related to temperature by $E_x = A + B\,T_x$. In the grey atmosphere approximation, $B \sim 4\sigma T^3/(1 + 0.75\tau)$, with τ being the infrared optical depth and σ being the Stefan-Boltzmann constant. Earth has an optical depth of about $\tau \sim 4$ and $T \sim 290\,\mathrm{K}$, so $B \sim 2\,\mathrm{W\,m^{-2}\,K^{-1}}$. (Lorenz et al. (2001) use a 'false' $\tau = 0.9$ in order for the grey radiative approximation to yield the correct surface temperature, but of course the atmosphere is actually in radiative-convective equilibrium, not in purely radiative equilibrium, and the real opacity is around 4. The real opacity yields an estimate for B that is closer to the empirical one from satellite observations.) If heat is transferred between the boxes at a rate $F = 2D(T_0 - T_1)$, then it is easy to show that the temperature difference ΔT is given by $\Delta T = T_0 - T_1 = (I_0 - I_1)/(B + 4D)$. For this formulation the entropy production dS/dT has a maximum for $D = B/4$, i.e., ~ 0.5–$0.8\,\mathrm{Wm^{-2}\,K^{-1}}$ – similar to observed values and leading to a temperature gradient consistent with observations (Fig. 12.1a). Somewhat equivalently, in the absence of atmospheric opacity effects, it emerges that the observed temperatures require D to have a numerical value close or equal to, the planet's average entropy production I/T.

12.1.2 Titan

Saturn's cold atmosphere-shrouded moon Titan has a thick, slowly-rotating atmosphere which exhibits a surprisingly large brightness temperature contrast of about 3 K compared with an equatorial temperature of about 93 K. Lorenz et al. (2001) noted that the observed brightness temperature contrast is exactly what would be predicted from MEP. Zonal models using mass-scaled D of 10^2 $-10^3\,\mathrm{W\,m^{-2}\,K^{-1}}$ predict contrasts of around 0.01 K. In contrast, the MEP principle mandates a D value rather lower than Earth of about $D \sim 4\sigma(93)^3/(1+0.75\tau)$ or around $0.02\,\mathrm{Wm^{-2}\,K^{-1}}$, leading to temperature contrast of a few K as observed (Fig. 12.1b). It is not clear how Titan's apparent heat transport is so inefficient. It may be that the latitudinal winds are suppressed by the strong zonal wind field, or that a condensation/evaporation phenomenon pins the polar temperatures at a low value. However, the *prima facie* agreement of the principle with the observed temperatures on Titan lends strong support to the MEP principle although some uncertainty regarding a possible stratospheric contribution to the brightness temperatures exists.

12.1.3 Mars

Lorenz et al. (2001) also consider the atmosphere of Mars. At first look, Mars does not obey MEP – its climate can be largely reproduced with the very small heat transport expected in a thin atmosphere (i.e., $D < 0.01\,\mathrm{W\,m^{-2}\,K^{-1}}$). However, its winter poles would get too cold with this value of D, and models are forced to pin them at the CO_2 condensation temperature of around 150 K. This is a quite reasonable 'fix' to the models, given that we can observe this process in action with the seasonal growth and

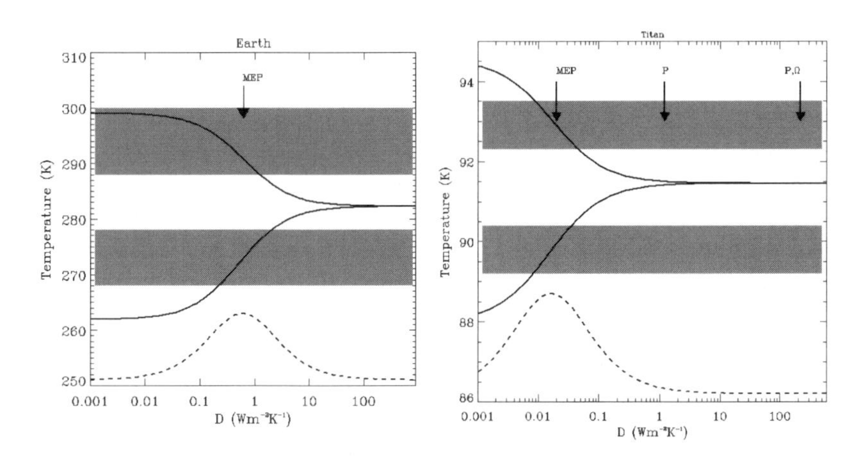

Fig. 12.1. Observed (*shaded boxes*) and modeled low and high latitude annual mean temperatures (*upper* and *lower lines* respectively). Simulated temperatures are a function of the heat transport parameter D. Entropy production (*dashed curve* – arbitary units) peaks for the Earth where the model curves agree with observations. On Titan, the entropy production curve again peaks (MEP) for the heat transport required to match observed temperatures, whereas pressure (P) and pressure/rotation rate scaling of the terrestrial values (P, Ω) fail, predicting temperature contrasts that are much too low. Adapted from Lorenz et al. (2001) and Ozawa et al. (2003)

decay of CO_2 frost caps. However, when the latent heat transport from the condensing cap to the atmosphere thence to the subliming cap is taken into account, the heat flow is entirely consistent with what would be predicted from MEP, of around $4\sigma(200)^3/(1+0.75\tau)$ or $\sim 0.4\,\mathrm{Wm^{-2}\,K^{-1}}$. Were meteorologists unaware of the CO_2 condensation, the predictions of GCMs or zonal models would be gravely in error, yet only knowledge of the basic radiative setting of the planet (insolation, obliquity, albedo and IR opacity) allows MEP to correctly predict that the atmosphere does *something* to ensure a heat transport of $\sim 0.4\,\mathrm{Wm^{-2}\,K^{-1}}$, corresponding to $\sim 1\,\mathrm{m}$ of seasonal frost and winds of several $\mathrm{ms^{-1}}$, as observed. This underscores the potential utility of MEP for astrobiological studies in predicting the resultant climate state, even when all the contributing mechanisms are not known.

12.1.4 Venus

On hot Venus, temperature contrasts are believed to be quite low although there are no accurate measures of surface temperature with latitude. Both, MEP and more conventional pressure-scaling of heat transport coefficients give more-or-less equivalent results. While Venus may indeed obey MEP, it is not an ideal test case. Furthermore, it may be that the altitudes at which solar flux is absorbed and re-emitted need to be considered as well.

12.1.5 Other Planets

The rapidly-rotating giant planets deserve further attention from a thermodynamic perspective. An additional degree of freedom that these bodies have is the significant heat flow from the interior, which can be of the same order as, or even exceed, the absorbed solar flux. The planet Jupiter, for example, has near-zero obliquity and therefore has a strong equator-to-pole insolation gradient, but the heat flux from the interior can be biased towards the polar regions in order to offset this gradient. Thus horizontal heat transport from low to high latitudes need not occur and indeed, the strongly belted structure of the atmosphere suggests that large-scale meridional motions in the troposphere are weak, as might be expected from the dynamical perspective of a large, rapidly-rotating planet. An additional question that exists for all planets, but has been largely ignored, is whether MEP applies to the surface, or some integrated absorption and emission of energy in the atmosphere, or only at some specific levels in the atmosphere. For the giant planets where there is no surface, this question becomes crucial. In passing, we should also note the information-theoretic application of maximum entropy principles to circulation on Jupiter by Sommeria (this volume).

12.1.6 Other Processes in Planetary Atmospheres

It may be that vertical convection in planetary atmospheres acts to extract work from the radiative setting in a very analogous way to the horizontal heat transport discussed earlier in this chapter. Ozawa and Ohmura (1997) showed how a simple 1-D radiative-convective model, where an initial radiative solution is perturbed by an arbitrary convective flux versus altitude, yields a vertical temperature profile very much like that observed on Earth. The magnitude of the convective flux is similarly consistent with observations. This approach is rather different in philosophy from, but perhaps more generally applicable than, the usual technique of pinning the lower atmosphere to some critical lapse rate, which is somewhat arbitrarily chosen. On Earth, the value most commonly observed is 6.5 K/km, intermediate between the dry adiabat of ~ 9.8 K/km and a moist adiabat of around 5 K/km. This 'convective adjustment' approach has proven popular, since by definition it yields the observed result on this planet. It might be argued, however, that the entropy production maximization technique suggested by Ozawa and Ohmura (1997) involves fewer ad-hoc assumptions.

Lorenz and McKay (2003) explore grey models with imposed lapse rates, and determine in the conventional way the resultant convective flux required to yield the imposed temperature profile. This is the conventional procedure, although the fluxes are rarely reported, particularly over such a wide range of opacities. They find that convection acts as if it 'short-circuits' a resistance to upward transport of heat (i.e., opacity). The convective flux – analogous to an electrical current through a motor wired across a radiative 'resistor' in an electrical circuit – varies as if the motor adjusted itself to maximize the electrical power dissipated within itself.

Finally, Verkley and Gerkema (2004) note that under the constraints of maximum entropy and constant integrated thermodynamic temperature, a vertical atmosphere in a gravitational field is isothermal, while under constraints of maximum entropy and constant integrated potential temperature, the isentropic (i.e., adiabatic) profile is obtained. Yet if all three constraints are applied simultaneously in a variational manner, the resulting profile has a weaker lapse rate than the dry adiabat, and in fact rather closely resembles the lapse rate observed on Earth.

12.2 A Probabilistic Explanation for MEP

A persistent difficulty with MEP has been the lack, until Dewar's work (2003, also this volume), of a persuasive rationale for why a system like the climate should choose an MEP state. This can be explained as follows. The work output of the system is governed by the combination of heat flow and the Carnot efficiency for that heat flow, and thus is a curve that asymptotes to zero at low and high ($\Delta I/2$) heat flows, with an intermediate maximum. For the system to be in steady state, the frictional dissipation must balance out the work production.

We may consider the circulation as the combination of many flow modes, each of which is characterized by a certain amount of heat transport (F_i), and a certain amount of loss by frictional dissipation (L_i). These modes may be very efficient in the sense of having low L/F – a large-scale ocean current, for example, or very inefficient (high L/F) like a small-scale eddy. The aggregate heat transport of the system is simply $F = \Sigma F_{\mathrm{i}}$, and the dissipation $L = \Sigma L_{\mathrm{i}}$. At steady state, work output and dissipation are balanced, and thus $L \leq F \, \Delta T/T$.

If we presuppose that the system has many modes available that can combine to satisfy these constraints, then it follows that there are many more possible microscopic combinations of modes that yield a macroscopic steady state when that steady state has a higher value of dissipation and work output (along the lines of Dewar's interpretation of MEP, see also related discussion on fluid turbulence in Sommeria, this volume, and spatial degrees of freedom in atmospheric turbulence in Ito and Kleidon, this volume). If all possible combinations are populated with equal probability, then the most likely states are those with higher dissipation.

If the system does not have sufficiently different modes available to attain MEP, for example if all L_i are so large as to exceed $F \, \Delta T/T$, then clearly MEP cannot be a steady state. The system would seize up. An example of this situation is a rapidly-rotating planet, where the dynamic constraints on fluid motion suppress large-scale eddies, forcing instead higher-vorticity, and thus more dissipative, modes. Similarly, a thin atmosphere with a low column mass cannot hold or transport much heat without requiring very high windspeeds and thus friction.

Hence MEP for a planetary climate is a probabilistic result – it is likely, but not guaranteed and subject to the dynamical constraints of the system. The framework above allows us to consider the often-expressed but vague condition for MEP of requiring 'sufficient degrees of freedom to choose a macroscopic MEP state'. Sufficient degrees of freedom for a system can be interpreted here as having sufficiently different modes such that combinatorial statistics favour the macroscopic states of maximum power.

12.3 Dissipation and Heat Transport

Besides the dissipation of energy by the motion in planetary atmospheres, there are also other important contributions to the entropy budget of a planet. While simulations of a dry atmosphere indicate that around 70% of the dissipation is due to mixing and the dilution of heat, and only 30% is due to frictional dissipation, recent studies by Pauluis and Held (2002a,b) describe how the dominant entropy source in the real terrestrial atmosphere is the irreversibility due to moist processes (see also Goody (2000); Pauluis, this volume). At the microscopic scale, evaporation leads to locally saturated air. Thus any air of relative humidity less than 100% implies that mixing has occurred, by diluting the locally saturated air just above the condensed phase with dry air from elsewhere. This mixing is irreversible in the sense that it leads to entropy generation.

In gross terms, the Martian atmosphere transports a global, annual average of some 25 W m^{-2} of heat. This heat is transported over a substantial temperature difference, which in principle should lead to substantial Carnot efficiencies $\Delta T/T$ of 25% , and thus significant mechanical work – indeed rather comparable with that performed by the Earth's atmosphere. And yet, observations of Martian sand dunes, for example, show that they do not move, whereas terrestrial sand dunes move at up to several tens of meters per year.

The paradox is then, how can the Martian atmosphere transport so much heat, yet do so without producing a commensurate amount of work? Unless purely mechanical considerations (cementing of the dunes, or some mechanical coupling inefficiency in the thin atmosphere) are responsible, the answer is likely to lie in the entropy generation associated with the Martian frost cycle and irreversibilities from the phase changes involved.

Certain small-scale 'dry' processes lack the entropy production associated with phase changes, and an idealized Carnot heat engine model can yield impressively accurate predictions: a notably successful application is the modeling of Martian dust devils (Rennò et al. 2000).

However, large-scale transports are dominated by the frost cycle, and thus the phase-change irreversibilities are likely to play an important role in limiting the efficiency of the Martian heat engine. Fig. 12.2 shows the instantaneous radiative entropy fluxes on Mars from a Global Circulation Model showing the important contribution of the frost cycle to the overall entropy production. Infrared observations from the Mars Global Surveyor satellite may permit a study of the martian entropy budget much as was done for the Earth (e.g., Stephens and O'Brien 1993) in the 1970s and subsequently from the Nimbus-7, ERBE and other satellites.

Conrath and Gierasch (1985) note another phase change responsible for entropy production and limitation of the efficiency of the atmospheric heat engine, namely the ortho:para hydrogen transition of molecular hydrogen, which is the dominant constituent of the outer planet atmospheres. The ortho:para ratio at thermodynamic equilibrium is a function of temperature, but has a finite relaxation time. Thus large circulations may, for example, bring gas from depth more quickly than the relaxation time and introduce disequilibrium concentrations of para hydrogen. Since the ortho:para ratio can be determined from the speed of sound, or remotely by infrared spectroscopy, this disequilibrium may serve as a useful tracer of vertical motions. Another disequilibrium process that leads to significant dissipation in the atmosphere is the frictional dissipation around falling raindrops (e.g., Lorenz and Rennò 2002; Pauluis, this volume).

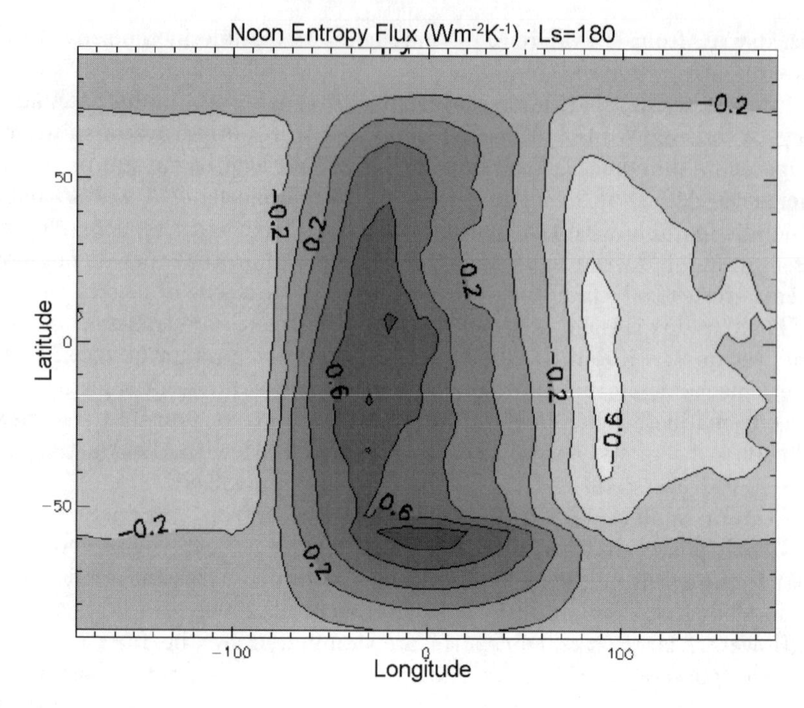

Fig. 12.2. Entropy Budget of Mars at L_s=180 using synthetic data from the European Mars Climate Database (available at http://www-mars.lmd.jussieu.fr/mars/access.html). Close to local noon (longitude 0), the budget is positive at around 0.6 $\mathrm{Wm^{-2}\,K^{-1}}$ as the surface is absorbing heat. It falls sharply in the early evening to a similar, but negative value, as the surface rejects heat it has accumulated during the day. Note that the southern frost cap (extending south of 60°S at this season) has a substantially positive entropy budget as radiates far less heat than it receives. Because its temperature is still pinned by CO2 frost, the cap edge has the highest entropy production of over 0.8 $\mathrm{Wm^{-2}\,K^{-1}}$ – the entropy budget has different information than simply the net flux

12.4 Geomorphology and Dissipative Structures

States of MEP can lead to highly organized, self-similar structures (see e.g., discussion on self-organized criticality and how it relates to MEP in Dewar, this volume). While some geomorphic structures on planets such as impact craters or volcanoes are the result of instantaneous catastrophic events, many other geological processes occur in settings where self-organization can occur: examples that have been suggested in the context of MEP are mantle convection (Vanyo and Paltridge 1981), sand dunes, and sorted stone circles (see also Ozawa et al. 2003).

Sand dunes are created by, and enhance, the exchange of momentum between the atmosphere and the ground. Their regular structures point to self-organization and Werner (1995) has noted that they are an 'attractor' –

initial distributions of sand over a surface in a simple model tend to converge to a state with dunes of a common form, dictated by the wind distribution and the sand supply. The hypothesis has been made that sand dunes may organize to optimize the sand transport normal to the dune crest. This concept may be directly analogous to one of Maximum Dissipation in that sand transport relates to the second or third power of shear velocity – the dune field may represent a sand system that has organized itself to maximally dissipate kinetic energy from the wind field while retaining a persistent organized form.

Another striking self-organized landform is the sorted circles that form in some frozen terrain. The circles, around a meter across, are formed from coarse stones that are segregated from the rest of the soil by repeated freeze-thaw cycles. (Kessler and Werner 2003) describe features in Alaska and Spitzbergen, where the observed morphology of the patterned ground (a labyrinth, or stripes, or circles etc.) depends on the stone:soil ratio, the surface slope and other parameters.

The role of entropy and thermodynamics in landscape processes was noted by Leopold and Langbein (1962): in particular the application of these ideas to river networks has received much attention in more recent years – see Miyamoto et al. (this volume) for more detailed discussion.

12.5 The Yarkovsky Effect – Migration of Meteorites via a Photon Heat Engine

In the 1990s, as both the Mars meteorite ALH84001 demanded attention to the migration of small bodies in space, and as telescopic surveys began to systematically inventory the population of small earth-crossing asteroids, a subtle effect has come to the fore in astrodynamics, now known as the Yarkovsky effect after a Polish engineer who discussed it around 1900 (but whose work has since been lost, see e.g., Farinella et al. 1998). This effect can be understood by noting that every photon has a tiny amount of momentum. If the photon is absorbed, or reflected, by a surface, then the surface must absorb, or reverse respectively, the momentum of the photon, and in so doing receives the momentum from it (or double, in the case of reflection). Thus a surface exposed to the sun experiences a momentum flux, a radiation pressure. This small force affects small particles with large area:mass ratios and is responsible for comet dust tails pointing away from the Sun and is the basis of 'solar sailing' (see also Burns et al. 1979).

But just as the short wavelength, high energy photons of sunlight exert a radiation pressure, so do the infrared photons associated with thermal emission, albeit less momentum per photon. Launching the photons exerts a small pressure, just as absorbing them does. So a hot surface experiences a small pressure, but a slightly larger pressure than a merely warm one. Yarkovsky realized that a body with an uneven temperature distribution would experience an uneven pressure, and hence a net force in space. A static body would of course be hottest on its sunlit side, and therefore experience a net force away from the sun due to the flux of thermal photons from that side. Consider an object in a circular orbit moving at right angles to the sun vector and with its axis of rotation orthogonal to the orbital plane (Fig. 12.3). If the body is rotating very slowly, the temperature has a strong noontime peak, exerting a radiation pressure away from the sun. But since the body is moving orthogonally to the radiation pressure force, the force performs no work on the body and its orbit does not change. On the other hand, if the body rotates quickly, the heat from the sun is smeared out over all angles, or, equivalently, all local times. The radiation pressure in all directions cancel out so there is no net force. At some intermediate rotation rate there is still a significant temperature bulge, but slewed round to the afternoon. There is thus still a net thrust, with a component along the object's direction of motion. Work is therefore done on the body, and its orbital energy increases (note the conceptual similarity with the climate model earlier). There are some variations on the Yarkovsky effect, depending on the eccentricity and inclination of the orbit, and the angle between the rotation axis of the asteroids and its orbital axis, but these are not be discussed here.

Lorenz and Spitale (2004) applied a thermodynamic analysis to the Yarkovsky effect. They noted that a simple linearized expression for the Yarkovsky force (e.g., Burns et al. 1979) may be written as $F_y = (8/3)\pi R^2 (\sigma T^4/\mathrm{c}) (\Delta T/T)$. Since the input power is $\sim 4\pi R^2 (\sigma T^4)$ and the work being done on the body moving at orbital speed v is vF_y it follows that the conventional engineering efficiency is $\sim (2/3)(\Delta T/T)\ (v/c)$. This can be seen as the product of a Carnot efficiency and a propulsive efficiency (v/c), a common term in rocketry, where the exhaust velocity of the propellant, here the speed of light c, should be matched to the flight speed v for optimum momentum transfer.

The Yarkovsky force is most significant for approximately meter to decameter scales – i.e., meteorites and asteroid fragments. The force is reduced from its 'ideal' value for small objects, where the distance scale is short enough for some of the dayside heat to be conducted through the object and radiated from the nightside (see e.g., Vokrouhlický 1998). Conventionally this may be viewed as a reduction in the thrust asymmetry, but from the thermodynamic perspective, it is a conductive heat loss that 'shorts out' the rotating heat engine.

It is reassuring that the Yarkovsky effect falls comfortably into the paradigm of a heat engine, although it is not yet clear how thermodynamics may offer new insights that the conventional momentum accounting approaches have not (although, see e.g., Fort et al. 1999 for an entropy treatment of stellar limb-darkening). Nonetheless, it is possible (e.g., Lorenz 2002) that interacting particles such as those in a planetary ring or circumstellar nebula may have degrees of freedom that permit self-organization and optimization of entropy production (see also Sommeria, this volume).

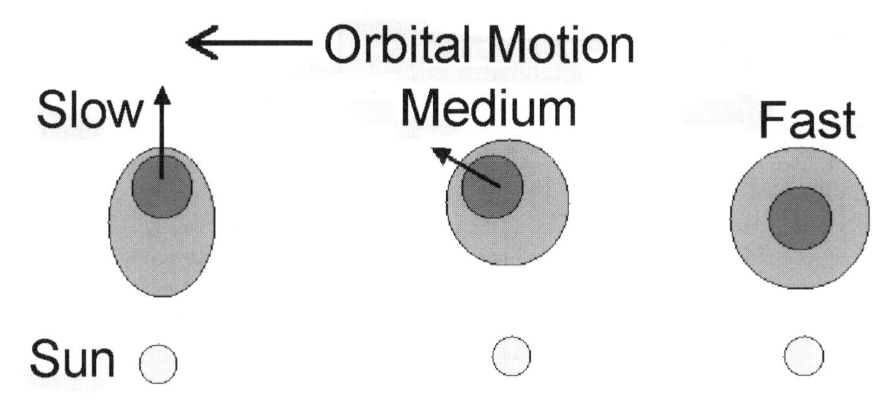

Fig. 12.3. Schematic diagram of the Yarkovsky effect for a spherical asteroid (mid-grey circle) with its rotation axis normal to the orbital plane. For a slow rotator (*left*) the temperature distribution (dark grey ellipse) is in equilibrium with sunlight, with a strong noontime temperature bulge: the thrust from the reradiated thermal radiation is therefore maximized, but is orthogonal to the direction of motion and thus performs no work. A fast rotator (*right*) has a near-isothermal temperature distribution and there is no net photon thrust. At an intermediate rotation rate, the maximum temperature occurs in the afternoon, such that the photon thrust, albeit slightly lower than the maximum at left, has a significant component along the direction of motion and thus the work output is maximized

12.6 Dyson Sphere – The Ultimate Stage in Planetary Evolution

The paradigm of a civilization's ultimate goal as the total downconversion of a star's light into thermal radiation is a neat image of life and order as being driven by thermodynamics and seeking to maximize their entropy production (see also Chaisson, this volume). Freeman Dyson suggested (Dyson 1960) that the presence of an advanced civilization might be revealed by the enhancement of infrared flux from a star. Mass and energy (or more specifically, available energy) are the limiting factors in growth, and Dyson noted that the growth in population and energy use of a civilization is much more rapid than the evolution of a star's luminosity, and thus a growing civilization would want to exploit an ever-larger fraction of the energy being radiated from its parent star. In particular, he notes that an expansion of population and industry of only 1 per cent per year for 3000 years would lead to an increase in energy use by a factor of 10^{12}. The human species presently exploits the mass of the biosphere of $5\ 10^{16}$ kg, consuming 10^{13} W. A factor of 10^{12} increase in these quantities corresponds roughly to the maximum mass reasonably accessible to humanity, taken as the mass of the planet Jupiter (2×10^{27} kg) and to the total energy output of the Sun, namely 4×10^{26} W. To exploit this energy output would require both a device for intercepting it (a shell to capture the sunlight) and the rejection of that energy at a lower temperature.

Dyson noted that the mass of Jupiter could be rearranged into a spherical shell at 2 astronomical units from the Sun. The thickness of this shell would be a few meters, depending on the density, and the shell would have an equilibrium temperature about equal to the Earth's present value (although some engineering details such as the heat transport across the shell have not been considered here). Well-behaved stars lie on an evolutionary track on the Hertzsprung-Russell diagram, a plot of luminosity with spectral type. If a star were to be obscured by a Dyson sphere, its emission spectrum would shift to the infrared, compared with stars of similar luminosity.

Although this idea remains interesting as a thought experiment, much recent astronomical work has been devoted to detecting such infrared excess not in order to search for extraterrestrial civilizations, but in order to detect non-artificial opaque envelopes around stars. Specifically, a cloud of dust in a protoplanetary nebula may surround a star, blocking its light but re-emitting its energy as infrared just as would a Dyson sphere. Thus infrared excesses are signals that something interesting is happening around a star, but is not necessarily a signature of astroengineering.

12.7 Concluding Remarks

This chapter surveyed some applications of nonequilibrium thermodynamics in the planetary sciences. Thermodynamic principles appear of particular utility in situations where there is little real data and thus the study of other planets can benefit from these ideas.

References

Aoki I (1983) Entropy productions on the earth and other planets of solar system, J. Phys. Soc. Jpn., 52, 1075–1078.

Burns JA, Lamy PL, Soter S (1979) Radiation forces on Small Particles in the Solar System, Icarus 40: 1–48.

Conrath B and Gierasch P (1985) Energy Conversion Processes in Outer Planet Atmospheres, in G E Hunt (ed) Recent Advances in Planetary Meteorology, Cambridge University Press, Cambridge, UK.

Dewar RC (2003) Information theory explanation of the fluctuation theorem, maximum entropy production, and self-organized criticality in non-equilibrium stationary states. J Physics A 36: 631–641.

Dyson FJ (1960) Search for Artificial Sources of Infrared Radiation. Science 131: 1667–1668.

Farinella P, Vokrouhlický D, Hartmann W (1998) Meteorite Delivery via Yarkovsky Orbital Drift. Icarus 132: 378–387.

Fort J, Gonzalez J, LLebot J, Saurina J (1999) Information theory and blackbody radiation. Contemporary Physics 40: 57–70.

Goody RM (2000) Sources and sinks of climate entropy. Q J Roy Meteorol Soc 126: 1953–1970.

Kessler MA, Werner BT (2003) Self-Organization of Sorted Patterned Ground. Science 299: 380–383.

Leopold LB, Langbein WB (1962) The Concept of Entropy in Landscape Evolution. U S Geological Survey Professional Paper 500-A.

Lorenz EN (1960) Generation of available potential energy and the intensity of the general circulation. in Pfeffer, RL (ed), Dynamics of Climate, Pergamon, Tarrytown, NY, pp 86–92.

Lorenz RD (2002) Planets, Life and the Production of Entropy. International Journal of Astrobiology 1: 1–13.

Lorenz RD, McKay CP (2003) A Simple Expression for Vertical Convective Fluxes in Planetary Atmospheres. Icarus 165: 407–413.

Lorenz RD, Rennó NO (2002) Work Output of Planetary Atmospheric Engines: Dissipation in Clouds and Rain. Geophys Res Lett 29: 1023, doi:10.1029/2001GL013771.

Lorenz RD, Spitale JN (2004) The Yarkovsky Effect as a Heat Engine. Icarus, 170:229–233.

Lorenz RD, Lunine JI, Withers PG, McKay CP (2001) Titan, Mars and Earth: Entropy production by latitudinal heat transport. Geophys Res Lett 28: 415–418.

Ozawa H, Ohmura A (1997) Thermodynamics of a global-mean state of the atmosphere – A state of maximum entropy increase. J Clim 10: 441–445.

Ozawa H, Ohmura A, Lorenz RD, Pujol T (2003) The Second Law of Thermodynamics and the Global Climate System – A Review. Rev Geophys 41: 1018. doi: 10.1029/2002RG000113.

Paltridge GW (1975) Global dynamics and climate – a system of minimum entropy exchange, Q J R Meteorol Soc 101: 475–484.

Pauluis O, Held IM (2002a) Entropy Budget of an Atmosphere in Radiative–Convective Equilibrium. Part II: Latent Heat Transport and Moist Processes. J Atm Sci 59: 140–150.

Pauluis O, Held IM (2002b) Entropy Budget of an Atmosphere in Radiative–Convective Equilibrium. Part I: Maximum Work and Frictional Dissipation. J Atm Sci 59: 125–139.

Rennò, NO Nash AA, Lunine JI, Murphy J (2000) Martian and terrestrial dust devils: Test of a scaling theory using Pathfinder data. J Geophys Res 105: 1859–1866.

Stephens GL, O'Brien DO (1993) Entropy and climate, I, ERBE observations of the entropy production. Q J Roy Meteorol Soc 119: 121–152.

Vanyo JP, Paltridge GW (1981) A model for energy dissipation at the mantle-core boundary. Geophys J R Astron Soc 66: 677–690.

Verkley, WTM, Gerkema, T (2002) Maximum entropy profiles. J Atm Sci 61:931–936.

Vokrouhlický D (1998) Diurnal Yarkovsky effect as a source of mobility of meter-sized asteroidal fragments. Astronomy and Astrophysics 335: 1093–1100.

Werner BT (1995) Eolian Dunes: Computer Simulations and Attractor Interpretation. Geology 23: 1107–1110.

13 The Free-Energy Transduction and Entropy Production in Initial Photosynthetic Reactions

Davor Juretić and Paško Županović

Faculty of Natural Sciences, Mathematics and Education, University of Split, Split, Croatia

Summary. Initial photosynthetic reactions in bacterial photosynthesis are modeled in this chapter as transitions among discrete states of integral membrane proteins performing these reactions. The steady state entropy production is then associated with each transition starting from photon absorption. The assumption of maximum entropy production in all irreversible non-slip transitions leads to high free energy transfer efficiency (high quantum yield) and to optimal values of kinetic constants that are comparable to experimentally determined values. Optimal overall efficiency of close to 20% is similar to measured values for the efficiency of producing the protonmotive power in reconstituted systems. We conclude that photosynthetic proton pumps operate close to maximum entropy production regime and use the advantage of nonlinear flux-force relationships to transfer power with around 90% efficiency instead of 50% as prescribed by the maximal power transfer theorem in the linear regime. Finally, the evolution-coupling hypothesis is suggested, according to which photoconverters couple their own evolution to thermodynamic evolution in a positive feedback loop accelerating both evolutions.

13.1 Introduction

The optimization and dissipation in initial photosynthetic reactions is very challenging problem, because even the most basic questions are still in dispute, such as can thermodynamics be applied at all (Hill 1977), and if it can, are equilibrium or near-equilibrium thermodynamics appropriate tools (Meszéna and Westerhoff 1999). Attempts to apply thermodynamics to photosynthesis mainly belong to two categories. In the first and earlier class are attempts to divine what was evolution's goal in photosynthesis. It is immediately obvious that maximal free-energy transduction efficiency was never evolution's goal, because of measured low efficiency values. More promising was the assumption that maximal output power must have been evolution's goal from the very start (Knox 1977; Lavergne and Joliot 2000). This idea worked only for very simple two-state photosynthetic models, when the requirement of maximal output power produced high optimal quantum yield in accord with spectroscopic measurements. For more complex models that allow for charge separation and the creation of a electrochemical proton gradient as the output force, it is far from clear that evolution tried to maximize the output power as the product of proton flux and protonmotive force.

Instead of guessing what evolution's goal is, some researchers speculated that photosynthesis must be governed by known principles from the thermodynamics of irreversible processes (Andriesse and Hollestelle 2001). Such an approach has the advantage of avoiding the artificial separation between the living and nonliving world, between the cell's environmental evolution in accord with thermodynamic laws and the apparently purposeful evolution of biological macromolecules. It is only natural to try to apply the well known minimal entropy production theorem associated with Prigogine's name (Prigogine 1967; see also Kleidon and Lorenz, this volume). Unfortunately, when applied to photosynthesis, authors missed the main point of that theorem that it describes a very special non-equilibrium state (the static head state), and performed their calculations assuming equilibrium (Andriesse 2000; Andriesse and Hollestelle 2001; Juretić 2002). The correct application of Prigogine's theorem in photosynthesis would not help, however, because the static head state is associated with zero net proton flux and accordingly vanishing free-energy transduction efficiency. Such a state can serve as the blockage to free-energy transduction, and is unlikely to describe the essence of photosynthesis. We shall also see that in the absence of leak the static head state of zero proton flux can never be reached in photosynthesis. No less serious is the objection to the application of Prigogine's theorem that this theorem is valid only for linear flux-force relationships close to equilibrium (Hunt and Hunt 1987). Measured flux-force relationships are nonlinear and both, input and output forces, are large keeping the system assuredly far from equilibrium (Cotton et al. 1984; Gräber et al. 1984; Wanders and Westerhoff 1988).

What is an alternative to minimum entropy production as the optimization principle? Is there any other "thermodynamic criterion" that would provide causal explanation for efficient free-energy transfer and conversion in bioenergetics? We propose in this work that the maximum entropy production principle (Dewar 2003; also Dewar, this volume) is relevant for photosynthesis. The maximum entropy production principle (MEP) has extremely wide applications (this volume). We applied it recently to standard electrical circuits, photovoltaic cells, biochemical circuits and photosynthetic free-energy conversions (Juretić and Županović 2003; Županović and Juretić 2004). The MEP principle can be applied both for linear and nonlinear flux-force relationships (Onsager 1931; Kohler 1948; Paltridge 1979; Juretić 1983,1984). In the linear range it is equivalent to Kirchhoff's laws for current distributions in the steady state for fixed parameters (Jeans 1923; Županović et al. 2004).

13.2 The Two-State Kinetic Model

The system at temperature T, illuminated with radiation at higher effective temperature T_R, can absorb maximal free energy A_{oc}, which is equal to the chemical potential of a photon (Meszéna and Westerhoff, 1999):

$$A_{oc} = h\nu(1 - T/T_R) \tag{13.1}$$

where $h\nu$ is the photon energy.

The steady state affinity A of a pigment P is (Lavergne and Joliot 1996):

$$A = h\nu + k_B T \ln([P^*]/[P]) \tag{13.2}$$

where $k_B$is the Boltzmann constant, and $[P]$ and $[P^*]$ are fractions of ground and excited chlorophyll states respectively. Energy utilization is possible only if a branched pathway exists through which part of photon energy can be invested into charge separation and stored into an electrochemical form. Assuming no branched pathway, which would decrease the $[P^*]$, A is equal to A_{oc}. The efficiency of free energy transduction then vanishes.

Fortunately, nature has designed photosynthesis so that charge separation through a branched pathway takes place with very high efficiency when other chlorophyll molecules are close by and a special chlorophyll molecule is strategically located close to an electron acceptor and an electron donor. As in an electrical circuit, when net electron current J flows, dissipation occurs and steady state affinity or photocell voltage (13.1) will be decreased. We know that the large majority of absorbed photons are converted into photelectrons, which implies that a photosynthetic system works far from the chemical equilibrium state for absorbed and emitted photons.

A convenient quantity to measure the distance from the chemical equilibrium state ($A = A_{oc}$) is the thermodynamic force for light reactions (Meszéna and Westerhoff 1999):

$$X_L = A_{oc} - A \geq 0 \tag{13.3}$$

The affinity transfer efficiency can then be defined as A/A_{oc}.

The entropy production P per the unit volume in the branched productive pathway due to transmitted free energy A and electron flux J is:

$$TP = AJ \tag{13.4}$$

where the electron flux J is proportional to the flux I of absorbed photons, but is an exponential function of the force X_L:

$$J = I(1 - \exp(-X_L/k_B T)) \tag{13.5}$$

where we assumed that the photon flux is approximately equal to the rate constant for photon absorption, because $[P^*] \ll [P]$.

For an optimal thermodynamic force for light transitions, or for corresponding optimal photochemical yield $\Phi = J/I$, entropy production is maximal. Maximal entropy production does not lead to poor affinity transfer efficiency or poor photochemical yield. In fact, due to the strongly nonlinear relationship (13.5) and the convexity of the graph of the current versus transmitted free-energy, maximum entropy production must occur both for high affinity transfer efficiency and for high photochemical yield. For assumed photon wavelength of 870 nm, an environment temperature of 25° C, non-radiative relaxation constant $k_d = 10^8\,\mathrm{s}^{-1}$, and light absorption rate $I = 100\,\mathrm{s}^{-1}$, maximal entropy production $P_{\max} = 29.5\,\mathrm{kJ\,mol^{-1}\,s^{-1}\,K^{-1}}$ is obtained for an optimal thermodynamic force for light transitions of $X_L = 9\,\mathrm{kJ/mol}$ (0.093 V), for an optimal photochemical yield $\Phi = 0.97$ and an optimal affinity transfer efficiency $A/A_{oc} = 0.91$. The experimentally observed photochemical yield values are indeed close to one (Cho et al. 1984; Lavergne and Joliot 1996). With optimal quantum yield close to one, only about 10% of available power is dissipated in the pathway $P \rightarrow P^*$ where energy utilisation cannot occur. In other words, the entropy production expression (13.4), which is associated with a potentially productive electron transfer pathway, is the major part of the total entropy production. Maximizing entropy production, in the case of a nonlinear current-force relationship, ensures that most of the absorbed power is channelled in the charge-separation pathway. Minimum entropy production would require working in the linear regime, when $\Phi \ll 1$ and $X_L \ll k_B T$, and having either zero affinity transfer efficiency, or zero photochemical yield. In the linear regime, in terms of the theory of electrical circuits, A_{oc} can be identified as the electromotive force, A as the voltage drop on a load, and X_L as the voltage drop through internal resistance. The entropy production (13.4) can be recognized then as the dissipation on the external resistor. It is maximal for the impedance matching requirement:

$$A_{optimal} = (X_L)_{optimal} = A_{oc}/2 \tag{13.6}$$

which is also known as the maximum power transfer theorem (Boylestad 1999). This requirement ensures that an optimal external resistance is equal to the internal resistance, so that 50% of the available source power is dissipated on the load.

13.3 The Five State Model for Chlorophyll Based Photoconversion

The scheme shown in Fig. 13.1 is a simplified five-state model for an-oxygenic chlorophyll-based bacterial photosynthesis (Van Rotterdam 1998; Lavergne and Joliot 1996). The five states are the chlorophyll ground state P and the chlorophyll excited states $P^*, B \equiv P^+B^-{}_A$, $H \equiv P^+H^-{}_A$ and $Q \equiv P^+Q^-{}_A$. The electron transport is assumed to be coupled to proton pumping in the recovery $B4$ transition. The photochemistry quantum yield Φ is the ratio of the $J(B4)$ flux and the flux of absorbed photons $J(L)$.

The application of Kirchhoff's junction rule leads to relationships among currents:

$$\begin{aligned} J(L) &= J(D) + J(B1) \\ J(B1) &= J(B2) \\ J(B2) &= J(S) + J(B3) \\ J(B3) &= J(B4) \end{aligned} \tag{13.7}$$

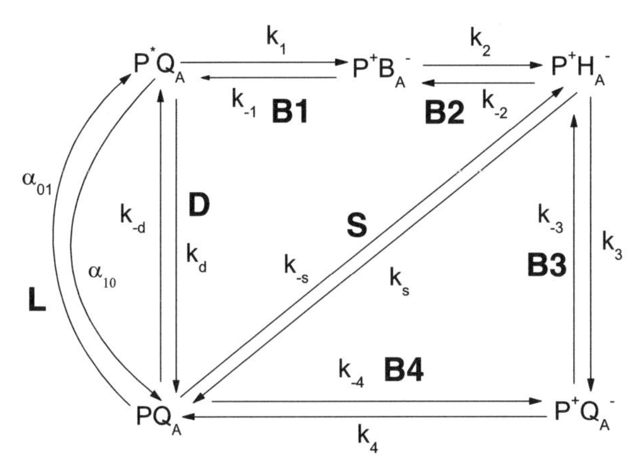

Fig. 13.1. The five-state kinetic model with a slip for the chlorophyll-based bacterial photosynthesis. The Q_A is ubiquinone electron acceptor, the B_A is accessory bacteriochlorophyll, while the H_A is pheophytin. Each transition is associated with forward and reverse rate constant. The excited state P^* is reached through a light-activated transition L between the chlorophyll ground (P) and the excited state, and depopulated through a non-radiative transition D back to ground state, and through relaxation $B1$ from the excited state with electron transfer and charge separation. Productive transitions leading to charge separation are $B1$, $B2$, $B3$ and $B4$

The affinity $A_{PP}*(L)$ is the thermodynamic force X_L introduced in (13.3) for the light-activated transition L. Kirchhoff's loop rule gives the connection between affinities and forces in each loop:

$$A_P *_P (D) = A_{oc} - X_L$$

$$A_P *_B (B1) + A_{BH}(B2) + A_{HP}(S) - A_P *_P (D) = 0 \tag{13.8}$$

$$A_{HQ}(B3) + A_{QP}(B4) - A_{HP}(S) = X_{out}$$

where input force A_{oc} and output force X_{out} can be derived by forming the clockwise and counterclockwise products of rate constants in cycles L-D and S-$B3$-$B4$ where these forces are respectively operational (Hill 1977).

The K_4 equilibrium constant for the $B4$ transition is a function of electron donor and acceptor concentrations and of the proton-motive force X_{out}:

$$K_4 = k_4/k_{-4} = \exp(u + X_{out}/k_B T) \tag{13.9}$$

where the donor/acceptor ratio is equal to $\exp(u)$. Relationships (13.8) are used to find the equilibrium constants $K(B3)$ and $K(S)$. Using the diagram technique (Hill 1977) one can derive the dependence of the transition and the operational fluxes and affinities on rate constants.

In terms of transition affinities and fluxes, total free-energy dissipation associated with our five-state diagram is:

$$PT = A_{PP*}(L)\,J(L) + A_{P*P}(D)\,J(D) + A_{P*B}(B1)\,J(B1) + A_{BH}(B2)\,J(B2) + A_{HQ}(B3)\,J(B3) + A_{QP}(B4)\,J(B4) + A_{HP}(S)\,J(S) \quad (13.10)$$

De Donder's Theorem (De Donder and Van Rysselberghe 1936) requires that the product of each affinity with corresponding flux is positive definite. Each term from the right hand side of (13.10) can be interpreted as the free-energy dissipation associated with corresponding transitions. Another way to consider total free-energy dissipation is to realise that it is equal to free-energy change of the input and output sources (Hill 1977):

$$PT = A_{oc}J(L) + X_{out}J(B4) \quad (13.11)$$

The corresponding efficiency expression is:

$$\eta = -X_{out}\,J(B4)/A_{oc}\,J(L) \quad (13.12)$$

Both initial transitions $B1$ and $B2$ are regarded as close to equilibrium, with equilibrium constants $K_1 = 4.8$ and $K_2 = 7.1$, respectively (Van Rotterdam 1998). One can ask if entropy production in irreversible transitions associated with productive pathway can always be maximized? Extensive modelling convinced us that connected maximums in the productive pathway will always occur, if we take care to perform free-energy transduction in the normal operating regime far from the static head state (Juretić and Županović 2003). We performed such optimization with respect to the rate constants in transitions $B3$ and $B4$ with the other rate and equilibrium constants taken as fixed, using observed values (Van Rotterdam 1998). Optimal final values of forward kinetic constants were stable with respect to the choice of initial values from 10^{-30} to 10^{30} and the iteration to the optimum was very fast, usually taking place during the first 10 steps. With a choice of $u = 12$, $K_4 = 100$, $\alpha_{01} = 100\,\mathrm{s}^{-1}$, $X_{out} = -18.55\,\mathrm{kJ/mol}$, an overall optimal efficiency of 17.7% is obtained (which is quite high compared to the maximal efficiency of 18.4%) and optimal forward rate constants are: $k_3(\text{optimal}) = 2.15 \times 10^9\,\mathrm{s}^{-1}$, $k_4(\text{optimal}) = 254\,\mathrm{s}^{-1}$. As an example, entropy production in the recovery transition ($B4$) is depicted in the Fig. 13.2, as the function of recovery rate constant k_4. The total entropy production of $19.78\,\mathrm{kJ\,mol^{-1}\,K^{-1}\,s^{-1}}$ is distributed among transitions L, D, S, $B1$, $B2$, $B3$, $B4$ (Fig. 13.1) as 2.92, 0.33, 0.80, 0.10, 0.02, 13.55 and $2.06\,\mathrm{kJ\,mol^{-1}\,K^{-1}\,s^{-1}}$ respectively. Notice the high contribution of irreversible transitions $B3$ and $B4$ (78.8%) to the total entropy production. The thermodynamic force for light reactions is $X_L(\text{optimal}) = 12.08\,\mathrm{kJ/mol}$, the affinity transfer efficiency is $(A_{oc} - X_L)/A_{oc}(\text{optimal}) = 87.8\%$, and the photochemical yield is Φ (optimal) = 94.6% . Optimal values for these performance parameters are nearly constant over wide range of light intensities, while the optimal proton current $J(B4)$ is proportional to the photon absorption rate. These optimized performance parameters are in the rough accord with experimental data (Juretić and Županović 2003).

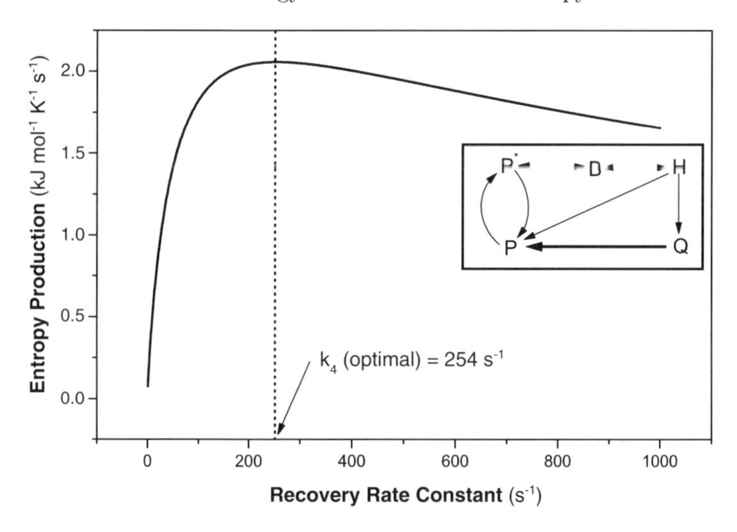

Fig. 13.2. The five-state kinetic model for chlorophyll-based photosynthesis has been optimized in the self-consistent manner so that entropy production in the chlorophyll recovery step is maximal in all non-slip irreversible dark transitions ($B3$ and $B4$ in Fig. 13.1). Insert: The bold line in the kinetic model specifies the recovery irreversible transition $B4$ for which the entropy production dependence on the forward rate constant k_4 is shown in the graph

13.4 Slip Coefficients and Forward Static Head State

The "forward" static head state (zero output flux) and corresponding slip coefficients have been defined for the nonlinear free-energy transduction as well (Juretić and Westerhoff 1987). We increased the secondary force by increasing the reverse recovery constant k_{-4} (Fig. 13.1). Total entropy production is low ($P = 2.3 \times 10^{-12}\,\mathrm{kJ\,mol^{-1}\,s^{-1}\,K^{-1}}$) in the state with vanishing proton flux, but it is still not minimal. The efficiency η is increased on approaching vanishing net proton flux, passes through maximum of almost 83% efficiency, and decreases to zero value.

The general expression for the static head output force is:

$$(X^{SH})_{out} = k_B T \ln((s_0 + 1)/(s_0 + \exp(X_{in}/k_B T))) \tag{13.13}$$

In the case of photosynthesis $X_{in} = A_{oc}$, where A_{oc} is given with (13.1). The forward slip coefficient s_0 is for the five-state chlorophyll based model:

$$s_0 = (k_d\, k_{-2}\, k_{-1} + k_s(k_1\, k_2 + (\alpha_{10} + k_d)(k_{-1} + k_2)))/(\alpha_{10}\, k_{-2}\, k_{-1}) \tag{13.14}$$

With rate constants and equilibrium constants for chlorophyll based photosynthesis taken from Van Rotterdam (1998) and as applied in our simplified kinetic model (Fig. 13.1), the secondary force high enough to cause vanishing flux $J(B4)$ is indeed very high (94.6 kJ/mol or 980 mV for $A_{oc} = 99$ kJ/mol and $s_0 = 5.7$). The static head state is actually very similar to an equilibrium state, because photon-activated cycling essentially stops and practically all chlorophyll is accumulated in the Q state due to a very low recovery equilibrium constant $K_4 = 8 \times 10^{-14}$. Of course, more complete kinetic models, e.g., including leak, can reach the steady state of zero net transmembane proton current for much lower (and more realistic) values of the proton-motive force.

13.5 Conclusions

When life learned how to use the photon free energy to perform charge separation, that major acomplishment steered all subsequent biological evolution. Charge separation leads, for instance, to the creation of the protonmotive force as the output force, which can be maintained only by continuous destruction of free energy packages. Maximal protonmotive force is associated with minimal (in the linear range) or very small entropy production (as demonstrated in this work for nonlinear flux-force relationships). However, net proton flux then vanishes, and ATP is not synthesized, so that uphill biosynthetic reactions, requiring ATP hydrolysis, cease to work. Since biological macromolecules are all more or less unstable and need to be constantly replaced, the cell is effectively dead in seconds after the ATP synthesis stops. Such a scenario never materializes due to the simple reason that the static head state with maximal protonmotive force cannot be reached. As soon as the membrane potential (the major part of the protonmotive force) reaches about 300 mV, the dielectric breakdown of the membrane occurs and the cell dies. Regulatory mechanisms maintaining the protonmotive force at a safe upper limit of around 200 mV must have been developed very early during life's evolution. Our point is that only a small percentage of photon free energy suffices to create such a protonmotive force, while the major part of incoming free energy packages must be dissipated. In other words the biochemical composition of cells is such that life is possible only far from thermodynamic equilibrium and far from the static head state. Then, entropy production is closer to maximal than minimal values (it is higher from the static head value by more than 12 orders of magnitude in our kinetic model for bacterial photosynthesis).

When all external forces are fixed (the photon free energy and protonmotive force) the conditions in the cell are analogous to an electrical network with all electromotive forces fixed. Then, Kirchhoff's laws and the condition of energy conservation requires that entropy production of the network is maximal (Jeans 1923; Ziman 1956; Županović et al. 2004). The analogues of Kirchhoff's laws and energy conservation condition holds for biochemical circuits as well (Županović and Juretić 2004). However, this theorem cannot be applied directly to biochemical circuits, because flux-force relationships in biochemistry are generally nonlinear and because biochemical circuits are not fixed in time, but rate constants and macromolecular states can change during evolution and even during the operation of the circuit. Taking advantage of the greater flexibility of biochemical networks we propose that all irreversible transitions involved in the power transfer are optimized with respect to rate constants so that the associated rates of entropy production are maximal. This proposal led to stable steady state with performance parameters, such as the efficiency of free energy transduction, close to experimentally measured values (Van Rotterdam et al. 2001). Using different kinetic models or different macromolecules for performing initial photosynthetic reactions (e.g., bacteriorhodopsin) does not change this conclusion (Juretić and Županović 2003). In addition, non-linearity turned out to be crucial, because optimal power transfer of around 90% is considerably higher than maximal power transfer of 50% in linear circuits. Due to the practically infinite source of free energy from the Sun, the advantage of the nonlinear mode in its superior capability to transfer, dissipate and store large amounts of free-energy is more important during evolution than its disadvantage in terms of limited overall efficiency, which is usually below 20% (Juretić 1992 and this work).

The proposed entropy production principle reverses the usual picture of what is important for a photosynthetic cell in its interaction with the environment. Its overall efficiency must be low enough to support a high level of entropy production associated with a rich pattern of metabolic fluxes. This may be quite general, because for most living cells, with photosynthethic ability or not, the major thermodynamic process is a large outflow of entropy, and less than 10% of available energy is incorporated into biomass (Bermudez and Wagensberg 1986). Notice that the steady state of exponential growth in a continuous culture experiments is preferred by microorganisms when all restrictions on growth factors have been removed (Forrest and Walker 1964). In such a steady state entropy change per unit mass of cells vanishes, but entropy production, entropy outflow and internal organization reaches its highest constant level. Living entities tend to increase the entropy production in the universe while active metabolically, so that life serves as a catalytic agent speeding entropy production in its environment (Ulanowicz and Hannon 1987; also Chaisson, this volume; Lineweaver, this volume). This observation brings biological evolution in synergy with the thermodynamic evolution. By operating close to maximal entropy production, photoconverters couple their own (biological) evolution to thermodynamic evolution in a positive feedback loop which speeds up both evolutions. This "evolution coupling" hypothesis postulates that biosphere evolution is intimately connected with the evolution of life's physical environment as suggested by the Gaia hypothesis (Lovelock and Margulis 1974; Pujol 2002; see also Kleidon and Fraedrich, this volume; Toniazzo et al., this volume). In this picture biological evolution is just a clever way nature found to accelerate its thermodynamic evolution. Life's particular goal seems to be to channel the input power into those dissipative pathways where electrochemical rather than only thermal free-energy conversions can occur.

References

Andriesse CD (2000) On the relation between stellar mass loss and luminosity. The Astronomical Journal 539: 364–365.

Andriesse CD, Hollestelle MJ (2001) Minimum entropy production in photosynthesis. Biophys Chem 90: 249–253.

Bermudez J, Wagensberg J (1986) On the entropy production in microbiological stationary states. J theor Biol 122: 347–358.

Boylestad R (1999) Introductory Circuit Analysis. Prentice-Hall, Upper Saddle River, NJ.

Cho HM, Mancino LJ, Blankenship RE (1984) Light saturation curves and quantum yields in reaction centers from photosynthethic bacteria. Biophys J 45: 455–461.

Cotton NP, Clark AJ, Jackson JB (1984) Changes in membrane ionic conductance, but not changes in slip, can account for the non-linear dependence of the electrochemical proton gradient upon the electron transport rate in chromatophores. Eur J Biochem 142: 193–198.

De Donder T, Van Rysselberghe P (1936) Thermodynamic Theory of Affinity. Stanford University Press, Stanford.

Dewar R (2003) Information theory explanation of the fluctation theorem, maximum entropy production and self-organized criticality in non-equilibrium stationary states. J Phys A 36: 631–641.

Forrest WW, Walker DJ (1964) Change in entropy during bacterial metabolism. Nature 201: 49–52.

Gräber P, Junesch U, Schatz GH (1984) Kinetics of proton-transport-coupled ATP synthesis in chloroplasts. Activation of the ATPase by an artificially generated ΔpH and $\Delta\Psi$. Ber Busenges Phys Chem 88: 599–608.

Hill TL (1977) Free Energy Transduction in Biology. The Steady State Kinetic and Thermodynamic Formalism. Academic Press, New York.

Hunt KLC, Hunt PM (1987) Dissipation in steady states of chemical systems and deviations from minimum entropy production. Physica 147A: 48–60.

Jeans JH (1923) The Mathematical Theory of Electricity and Magnetism. 4th edn. Cambridge Univerity Press, Cambridge.

Juretić D (1983) The thermodynamic and kinetic limits on the process of free energy storage by photosynthetic systems. Croatica Chemica Acta 56: 383–387.

Juretić D (1984) Efficiency of free energy transfer and entropy production in photosynthetic systems. J theor Biol 106: 315–327.

Juretić D, Westerhoff HV (1987) Variation of efficiency with free-energy dissipation in models of biological energy transduction. Biophys Chem 28: 21–34.

Juretić D (1992) Membrane free-energy converters: The benefits of intrinsic uncoupling and non-linearity. Acta Pharmaceutica 42: 373–376.

Juretić D (2002) Comment on 'Minimum entropy production in photosynthesis'. BioComplexity. http://mapmf.pmfst.hr/~juretic/Juretic-revised-comment.pdf

Juretić D, Županović P (2003) Photosynthetic models with maximum entropy production in irreversible charge transfer steps. Comp Biol Chem 27: 541–553.

Knox RS (1977) Photosynthetic efficiency and exciton transfer and trapping. In: Barber J(ed) Primary Processes in Photosynthesis, vol. 2, Elsevier, New York, pp 55–97.

Kohler M (1948) Behandlung von Nichtgleichgewichtsvorgängen mit Hilfe eines Extremalprinzips. Z Physik 124: 772–789.

Lovelock JE, Margulis L (1974) Atmospheric homeostasis by and for the biosphere: the gaia hypothesis. Tellus 36: 1–9.

Lavergne J, Joliot P (1996) Dissipation in bioenergetic electron transfer chains. Photosynthesis Research 48: 127–138.

Lavergne J, Joliot P (2000) Thermodynamics of the excited states of photosynthesis. BTOL-Bioenergetics (http://www.biophysics.org/btol/bioenerg.html)

Meszéna G, Westerhoff HV (1999) Non-equilibrium thermodynamics of light absorption. J Phys A 32: 301–311.

Onsager L (1931) Reciprocal relations in irreversible processes I. Phys Rev 37: 405–426.

Paltridge GW (1979) Climate and thermodynamic systems of maximum dissipation. Nature 279: 630–631.

Prigogine I (1967) Thermodynamics of irreversible processses. Interscience, New York.

Pujol T (2002) The consequence of maximum thermodynamic efficiency in Daisyworld. J theor Biol 217: 53–60.

Ulanowicz RE, Hannon B (1987) Life and the production of entropy. Proc R Soc Lond B 232: 181–192.

Van Rotterdam BJ (1998) Control of Light-Induced Electron Transfer in Bacterial Photosynthesis. Ph.D. thesis, University of Amsterdam.

Van Rotterdam BJ, Westerhoff HV, Visschers RW, Bloch DA, Hellingwerf KJ, Jones MR, Crielaard W (2001) Pumping capacity of bacterial reaction centers and backpressure regulation of energy transduction. Eur J Biochem 268: 958–970.

Wanders RJ, Westerhoff HV (1988) Sigmoidal relation between mitochondrial respiration and $\log([ATP]/[ADP])_{out}$ under conditions of extramitochondrial ATP utilization. Implications for the control and thermodynamics of oxidative phosphorylation. Biochemistry 27: 7832–7840.

Ziman JM (1956) The general variational principle of transport theory. Canadian Journal of Physics 34: 1256–1273.

Županović P, Juretić D (2004) The chemical cycle kinetics close to the equilibrium state and electrical circuit analogy. Croatica Chemica Acta, in press.

Županović P, Juretić D and Botrić S (2004) Kirchhoff's Loop Law and the principle of maximum entropy production. Phys Rev E, in press.

14 Biotic Entropy Production and Global Atmosphere-Biosphere Interactions

Axel Kleidon[1] and Klaus Fraedrich[2]

[1] Department of Geography and Earth System Science Interdisciplinary Center, 2181 Lefrak Hall, University of Maryland, College Park, MD 20742, USA
[2] Meteorologisches Institut, Universität Hamburg, Bundesstraße 55, 20146 Hamburg, Germany

Summary. Atmospheric conditions constrain biotic activity through incoming solar radiation, temperature, and soil water availability on land. At the same time atmospheric composition and the partitioning of energy fluxes at the surface are strongly affected by biotic activity, thereby modifying the environmental constraints. Here we review the foundations for atmosphere-biosphere interactions, focusing on the role of biogeophysical effects of terrestrial vegetation and the emergent feedbacks of the coupled atmosphere-biosphere system. We then investigate atmosphere-biosphere interactions from a perspective of entropy production and discuss the applicability of the hypothesis of Maximum Entropy Production (MEP) to biotic activity as a dissipative process within the Earth system. Specifically, we suggest two examples demonstrating the existence of MEP states associated with biotic activity and the Earth's planetary albedo. We close with a discussion of how this research can be extended and what the implications of biotic MEP states would be for understanding the dynamics of the Earth system.

14.1 Introduction

The biota plays an important role in the climate system. For instance, the strength of the atmospheric greenhouse, in terms of atmospheric concentrations of carbon dioxide, is closely linked with the global carbon cycle, which is affected by biotic activity through photosynthesis, respiration, and enhancement of rock weathering on geologic time scales (see also Schwartzman and Lineweaver, this volume). Over land, the absorption of solar radiation and the subsequent partitioning into radiative and turbulent heat fluxes is strongly affected by terrestrial vegetation. For instance, the presence of a rainforest leads to a darker and aerodynamically rougher surface with a higher capacity to evaporate water. These aspects affect the physical functioning of the land surface. However, biotic processes such as photosynthesis and respiration are also strongly constrained by the atmospheric environment, through temperature and the availability of light and water. This leads to the notion of atmosphere-biosphere interactions, with atmospheric conditions constraining biotic activity but biotic effects moderating atmospheric conditions, resulting in emergent feedbacks. In addition, both, an abiotic surface as well as a biotically influenced surface are consistent with the constraints imposed by the

Table 14.1. Albedoes for different surfaces and cloud types (after Hartmann 1994)

	typical value (in %)
bare surfaces:	
water	7–12
moist soil	10–25
dry soil	30–35
vegetated surfaces:	
short green vegetation	17
Extratropical forests	10–15
Tropical forests	13–18
snow- and ice-covered surfaces:	
sea ice, no snow cover	30
old, melting snow	50
fresh, dry snow	80
forest with surface snow cover	25
clouds:	
high (cirrus)	21
medium (cumulus)	48
low (stratus)	69

surface energy- and water balances, so that there are potentially many possible states and we may ask which biotic state is the most probable state. This then relates to the question of how atmosphere-biosphere interactions affect the rate of entropy production, and how the MEP principle can be applied to understand the emergent outcome of atmosphere-biosphere interactions.

Let us first briefly motivate the reasoning why biotic activity should affect the rate of entropy production at the planetary scale. Consider the planetary rate of entropy production, as approximated by

$$\sigma_{EARTH} = I_0(1 - \alpha_P)(1/T_R - 1/T_{SUN}) \qquad (14.1)$$

with I_0 being the mean solar irradiation at the top of the atmosphere, α_P the planetary albedo, T_R the effective radiative temperature of Earth, and T_{SUN} the emission temperature of solar radiation. The radiative temperature T_R is mainly constrained by the planetary energy balance, since absorption of solar radiation balances the emission of terrestrial radiation at the effective radiative temperature T_R in steady state. The planetary albedo plays a crucial role in determining σ_{EARTH}. It is not a fixed property, but results from the dynamics of the energy- and water balances, specifically the amount and type of clouds, the type of vegetative cover on land, and the abundance of snow or ice at the surface or the lack thereof (Table 14.1). Considering that many processes affect the planetary albedo, we may argue that many degrees of freedom are associated with total absorption of solar radiation, and that the Maximum Entropy Production (MEP) principle should be applicable.

Noting that vegetated surfaces are generally cooler than non-vegetated surface, Ulanowicz and Hannon (1987) and Schneider and Kay (1994) both argued that biotic influences on physical processes make them more efficient (by lowering surface temperature), resulting in higher rates of entropy production. Given that the biosphere is inherently diverse, with differing functional responses to environmental conditions of a large number of individuals, we may take their perspectives further and ask how the MEP principle should be applicable to atmosphere-biosphere interactions at the global scale.

In this chapter, we first review the basics of photosynthetic activity as the main energy source of the biota and how it is constrained by the climatic conditions at the surface. We then discuss the consequences of photosynthetic activity for energy- and water partitioning and how these effects impact the rate of biotic entropy production. Finally, we propose how *MEP* may be applicable to biotic activity in an Earth system context and how the emergent behavior may share similarities with the Gaia hypothesis of Lovelock and Margulis (1974) (also Toniazzo et al., this volume).

14.2 Photosynthetic Activity and Climatic Constraints

The majority of biotic activity results from photosynthesis and the derived organic carbon compounds. The assimilates from photosynthesis are used to maintain existing living tissues or allocated to growth and reproduction. Eventually, organic carbon compounds are respired – either by the photosynthesizers through autotrophic respiration, or by heterotrophic respiration at various trophic levels of the food chain. In the following we focus on the climatic constraints on photosynthesis as the primary process supplying energy to the biota.

14.2.1 Climatic Constraints on Biotic Productivity

The net conversion of carbon dioxide into organic carbon by photosynthesis can be expressed as

$$CO_2 + H_2O + \text{energy} \rightarrow HCO_2 + O \qquad (14.2)$$

The conversion of one mol of CO_2 into organic carbon requires roughly 479 kJ of energy (Larcher 1995). This energy is derived from the absorption of solar radiation by chlorophyll, particularly wavelengths between 380 nm and 710 nm. This band of radiation is also referred to as photosynthetically active radiation, or PAR. PAR represents roughly 50% of the incoming solar radiation, I_S. The amount of absorbed PAR, or $APAR$, depends on the photosynthesizing biomass, C_{GREEN}, and the efficiency of its conversion into organic carbon depends on limiting factors imposed by the availability of nutrients and the temperature of the environment. A large fraction of up to 75% for tropical rainforests (Larcher 1995) of the resulting gross uptake of carbon by photosynthesis (or gross primary productivity, GPP) is consumed by autotrophic respiration (RES_A). Net primary production, or NPP, describes the net carbon uptake by photosynthesizers that is not consumed by autotrophic respiration (i.e., $NPP = GPP - RES_A$).

NPP is commonly used to describe biotic activity. It can be estimated from the amount of $APAR$ by (Monsi and Saeki 1953; Monteith 1977; Field et al. 1998)

$$NPP = \varepsilon APAR \tag{14.3}$$

where ε is the average light use efficiency. The light use efficiency includes the environmental constraints on biotic activity imposed by temperature, and water and nutrient availability. By explicitly expressing the climatic effects on NPP through temperature and water availability, (14.3) can be expanded to:

$$NPP = \varepsilon_{MAX}\, f(C_{GREEN})\, g(T_S)\, h(W_S) PAR \tag{14.4}$$

where ε_{MAX} is the maximum light use efficiency (reflecting the biochemical efficiency of the conversion and other limitations not considered here, e.g., by nitrogen or phosphorus availability), W_S is the available soil moisture within the rooting zone (for terrestrial vegetation) and f, g, h some functional relationships.

Oceanic productivity is primarily limited by the availability of nutrients. Nutrients originate from deep ocean water where dead organic material is decomposed. Consequently, regions in which surface water mixes with deep water show a high productivity, such as the upwelling regions at the western shores of continents and in regions where the depth of the mixed layer shows large seasonal variations (as in the mid-latitudes), which allows for mixing with deeper water.

On land, geographic variations in productivity can be categorized in water-limited environments - mainly in the tropics – and in temperature-limited environments – mainly in temperate and polar regions. Water limitation to NPP arises from the uptake of atmospheric carbon dioxide through the plant's stomata being strongly coupled to the loss of water by transpiration. Consequently, productivity decreases with a decrease in precipitation, shaping the transition of vegetation types from rainforests to savanna to desert along moisture gradients. In polar regions, temperature limits the rate of photosynthesis, and the strong seasonality in solar radiation leads to a limited length of the growing season. This limitation is reflected in the transition of vegetation types of lower productivity towards colder regions, from temperate forests to boreal forest to tundra.

14.2.2 Dynamic Constraints of Terrestrial Energy- and Water Exchange

The temperature- and water constraints of terrestrial productivity are governed by the surface energy and water balance. The surface energy balance links the net radiative fluxes of solar and terrestrial radiation with the turbulent fluxes of sensible and latent heat. The net radiative energy flux, R_N,available for partitioning is the sum of the absorption of short wave radiation from the sun at the surface, I_S, reduced by the net emission of infrared radiation, $Q_{LW}(T_S)$, depending on the surface temperature, T_S, and the strength of the atmospheric greenhouse effect:

$$R_N = I_S(1 - \alpha_S) - Q_{LW}(T_S) \qquad (14.5)$$

with α_S being the albedo of the surface. This net radiative flux is partitioned into the sensible and latent heat fluxes, SH and LH, and the heat flux heating the ground, $c\,dT_S/dt$:

$$R_N = SH + LH + c\,dT_S/dt \qquad (14.6)$$

where c is the effective heat capacity of the soil. The chemical energy produced by photosynthesis and released by respiration is comparably small and is neglected here. If water availability is not limiting the latent heat flux (as is the case over the oceans) then the Bowen ratio, defined as $B = SH/LH$, can be approximated by the equilibrium Bowen ratio, B_e. The equilibrium Bowen ratio sets a fixed proportion of the two heat fluxes depending on surface temperature, with the partitioning shifting towards increased latent heat flux with increasing surface temperature. It is given by (Hartmann 1994):

$$B_e^{-1} = \lambda/c_p \; \partial q_{\mathrm{SAT}}/\partial T \; \approx \lambda^2/(c_p R T_S^2) \; q_{SAT}(T_S) \qquad (14.7)$$

where λ being the latent heat of vaporization, c_p the specific heat capacity of the air, R the gas constant for water vapor, and q_{SAT} the saturated specific humidity, which increases roughly exponentially with T_S.

Over land, the latent heat flux is often limited by the amount of available water, so that the equilibrium Bowen ratio only serves as an upper limit for the latent heat flux. The surface water budget describes another partitioning taking place at the surface and imposes an important constraint on the latent heat flux. The incoming flux of water (that is precipitation P and snowmelt SM) is partitioned into the fast surface and slow drainage runoff R, the evapotranspiration ET (with latent heat flux $LH = \lambda ET$) and the change in soil water storage within the rooting zone dW_S/dt:

$$P + SM = ET + R + dW_S/dt \qquad (14.8)$$

Equations (14.5), (14.6) and (14.8) represent the dynamical formulation of the constraints on terrestrial NPP. These constraints, however, are not fixed, but affected by the biota, leading interacting dynamics of the atmosphere-biosphere system and emergent feedbacks.

14.3 Biogeophysical Effects and Feedbacks

14.3.1 Vegetation Effects on Land Surface Characteristics

Biotic activity affects the physical exchanges of energy and water primarily at the land surface. Terrestrial vegetation modifies the physical aspects of land surface exchange by a series of characteristics (e.g., Kleidon et al. 2000; Bonan 2002; Pitman 2003):

- *Surface albedo:* vegetated surfaces are generally darker than non-vegetated surfaces, leading to increased absorption of incoming solar radiation at the surface (Table 14.1).
- *Albedo of snow cover:* the vertical structure of forests, particular coniferous forests in the boreal regions, leads to a masking effect of surface snow cover, effectively reducing the overall surface albedo in the presence of snow cover (Table 14.1). This leads to enhanced absorption of solar radiation, particularly during spring time, accelerating the rate of snow melt.
- *Surface roughness:* vegetated surfaces are aerodynamically rougher because of their heterogeneous canopy structure, leading to a shift in the energy balance towards greater turbulent fluxes of sensible and latent heat, reducing net radiative loss by terrestrial radiation.
- *Bowen ratio:* vegetated surfaces generally show higher fluxes of LH and reduced SH because of the direct coupling of carbon uptake with transpiration by stomatal functioning, leading to a lower Bowen ratio.
- *Rooting zone depth:* vegetation can more effectively take up soil moisture for transpiration through the vertical extent of the rooting zone, therefore enhancing the ability of a land surface to maintain ET during dry episodes, with further consequences on energy partitioning and surface temperature.

The magnitude of these effects on climate, and therefore on the climatic constraints for terrestrial productivity, can be estimated by extreme climate model simulations of a "Desert World" and a "Green Planet" (Fraedrich et al. 1999; Kleidon et al. 2000). In the "Desert World" simulation, land surface characteristics representative of a desert (i.e., high albedo, low roughness, low ability to evaporate water) were prescribed to all land surfaces with no permanent ice cover. Compared to the simulated climate of the "Present-Day", the surface receives less solar radiation in the "Desert World" climate, the surface energy balance is shifted towards more loss by terrestrial radiation and reduced rates of latent heat flux (Table 14.2). As a consequence of less evapotranspiration, precipitation over land is reduced by more than 30% . At the other extreme of a "Green Planet", land surface characteristics were prescribed to be representative of a rainforest (i.e., low albedo, high roughness, high ability to transpire water). When compared to the "Present-Day" climate, the simulated climate of the "Green Planet" shows the same differences, but of opposite sign: the surface energy balance is shifted towards more turbulent fluxes and evapotranspiration. Consequently, precipitation is enhanced by 30% . Net solar radiation at the surface, however, is slightly reduced as a consequence of increased cloud cover. Note that the land surface parameters in the "Green Planet" simulation do not reflect carbon constraints. That is, the land surface characteristics of a rainforest would not be sustained by the productivity of the vegetation in all areas of the simulated climate of the "Green Planet". On the other hand, some regions of the "Desert World" scenario would exhibit climates that would sustain vegetation, and it has been shown that the present-day vegetation-climate state is likely a reproducible state independent of initial conditions (Claussen 1994; Claussen 1998).

Table 14.2. Components of the surface energy budget, water cycle and terrestrial productivity averaged over land for a "Desert World", the "Present-Day", and the "Green Planet" (after Roeckner et al. 1996; Fraedrich et al. 1999; Kleidon et al. 2000; Kleidon 2002). Negative components represent net loss of energy from the surface. Terrestrial productivity has been normalized to yield 100% for the "Desert World" climatic conditions

	"Desert World"	"Present-Day"	"Green Planet"
Energy balance: (in W/m^2)			
net solar radiation	124	130	129
net terrestrial emission	-74	-62	-53
sensible heat flux	-22	-17	-8
latent heat flux	-18	-44	-60
Water cycle:			
precipitation (in 10^{12} m^3/yr)	71	108	137
evapotranspiration (in 10^{12} m^3/yr)	31	73	108
precipitable water (in kg/m^2)	16	18	21
cloud cover (in %)	51	53	58
Terrestrial productivity: (in %)	100	250	255

14.3.2 Climate Feedbacks of Terrestrial Vegetation

When the simulated climates are used to calculate terrestrial productivity from climatic constraints (following the approach in Sect. 14.2), the "Present-Day" climate allows for a terrestrial productivity 2.5 times the one corresponding to the "Desert World" climatic conditions (Table 14.2). This suggests that the overall feedback associated with biogeophysical effects is positive, that is, that the changes in the simulated climates that result from the inclusion of biotic effects lead to less climatic constraints and allow for a higher productivity (Betts 1999; Kleidon et al. 2000; Kleidon 2002). Note that the difference in productivity for the "Present Day" and the "Green Planet" climate is only marginal despite considerable differences in the simulated climates. This is due to the fact that the increased productivity in marginal areas is offset by the reduction in productivity in highly productive areas due to increased cloud cover.

The effects of terrestrial vegetation can be understood in terms of two *biogeographical* feedbacks (i.e., changes in land surface characteristics that are usually associated with changes in vegetation type which take place on time scales of decades to centuries):

- *boreal forest feedback:* In temperature-limited environments, the presence of forest leads to a lower surface albedo in the presence of snow (Table 14.1). This leads to the following feedback loop which reinforces the presence of forest (Bonan et al. 1992): + forest → − surface albedo (if snow is present); + absorption of solar radiation → + temperature; + snow melt → + length of growing season → + productivity to sustain forest. Ultimately, the amount of snowfall and the seasonality in solar radiation set an upper limit to the strength of this feedback. Climate model simulations have also shown that this feedback is amplified by changes in sea-ice at the hemispheric scale.
- *water cycling feedback:* In water-limited environments, the presence of vegetation allows for a lower surface albedo and better access soil moisture through a root system, both of which act to enhance evapotranspiration. The presence of vegetation is reinforced by the following feedback (Charney 1975; Milly and Dunne 1994; Eltahir 1998; Kleidon and Heimann 2000): + vegetation → + evapotranspiration/latent heat flux; − sensible heat flux → + water vapor in the planetary boundary layer; − boundary layer growth → + precipitation → + water availability → + productivity to sustain vegetation. Additional physical feedbacks lead to enhanced net radiation at the surface, reinforcing increased evapotranspiration (Charney 1975; Eltahir 1998, see also Table 14.2): + evapotranspiration → + water vapor in the planetary boundary layer → + enhanced absorption of longwave radiation → + incoming longwave radiation at the surface → + net radiation at the surface → + available energy for evapotranspiration; and: + evapotranspiration → − surface temperature → − emission of terrestrial radiation → + net radiation at the surface → + available energy for evapotranspiration. Negative feedbacks that set limits to these loops are through decreased water vapor pressure deficit in the planetary boundary layer, therefore reducing the atmospheric demand for evapotranspiration, and increased cloud cover which reduces incoming surface solar radiation (see also Table 14.2).

Both of these feedback loops extend the boundary of vegetation types towards the more limited side, therefore extending the overall area available for biotic productivity.

These feedbacks have been suggested to be important for understanding the climate system response to global change, for instance during glacial-interglacial cycles. During the Mid-Holocene (approx. 9000 years before present), the time that the Earth was closest to the sun on its orbit (i.e., perihelion) was during the northern hemisphere summer, so that the northern hemisphere received more solar radiation in summer. Foley et al. (1994) demonstrated that the northern shift of the boreal forest zone during that time was amplified by the boreal forest feedback, resulting in a simulated climate which is in better agreement with paleo-reconstructions. During the same period, reconstructions suggest that the Sahara desert was much reduced in size (as for instance depicted by cave paintings in the Sahara, leading to the notion of a "green" Sahara with abundant wildlife). Kutzbach et al. (1996) and Claussen and Gayler (1997) demonstrated that this reconstruction is reproduced with climate model simulations if the water recycling feedback associated with the shift of vegetation zones is included in the model. The boreal forest feedback has also been suggested as a potentially important factor for initiating ice ages (deNoblet et al. 1996; Gallimore and Kutzbach 1996), while the water recycling feedback (associated with deep rooted vegetation) has been suggested to be important for sustaining Amazonian rainforest cover during the last glacial maximum (Kleidon and Lorenz 2001).

14.4 Biotic Entropy Production and MEP

Biotic activity, mainly represented by photosynthesis and subsequent respiration of organic carbon, allows organisms to perform work and therefore leads to entropy production. Photosynthesis absorbs and utilizes a certain fraction of incoming low entropy solar radiation, that is, it converts solar energy into carbohydrates at a certain rate, Q_{GPP}. These carbohydrates are eventually respired, releasing carbon dioxide and heat roughly at the surface temperature T_S of the Earth. In steady state, the conversion of solar energy into carbohydrates by photosynthesis balances the production of heat by respiration (neglecting the effect of carbon burial), leading to a rate of biotic entropy production of (Kleidon, 2004a):

$$\sigma_{BIO} \approx Q_{GPP}(1/T_S - 1/T_{SUN}) \qquad (14.9)$$

Equation (14.9) determines the overall biotic entropy production from the differences of energy fluxes at the biotz-environment boundary, rather than summing up all factors that lead to entropy production within the biotz. With this expression we also neglect contributions of other metabolisms that are not related in the processing of carbohydrates derived from photosynthesis.

Previous work by Ulanowicz and Hannon (1987) and Schneider and Kay (1994) suggests that terrestrial vegetation acts to enhance the rate of entropy production by lowering the surface albedo and surface temperature. Here we suggest that the biota does not only increase the rate of entropy production, but that there are distinct macroscopic states of maximum entropy production associated with biotic activity.

14.4.1 Conditions for Biotic MEP States

As discussed in the introduction to the book (Kleidon and Lorenz, this volume), MEP applies to open thermodynamic systems which (a) do not have fixed boundary conditions and (b) have sufficient degrees of freedom (see also Dewar 2003, and Dewar, this volume). Biotic entropy production depends primarily on biotic productivity Q_{GPP}, which in turn depends on incoming solar radiation, water availability and surface temperature (in a similar way as expressed in equation (14.4)). The extreme climate model simulations discussed in the previous section illustrate that none of these factors is fixed, but strongly affected by the presence of terrestrial vegetation (Table 14.2). Consequently, biotic productivity on land is indeed subject to open boundary conditions. Furthermore, the terrestrial biota is inherently diverse, that is, there are many different ways for individual organisms to use assimilated carbon to grow, reproduce, and respond to environmental conditions (e.g., with respect to stomatal functioning, Buckley et al. 1999). Kleidon and Mooney (2000) used an individual-based modeling approach to demonstrate this functional diversity for terrestrial vegetation (Fig. 14.1). They developed a model of an individual plant, which simulates the growth and phenology as a function of environmental conditions, and then used it in a Monte-Carlo setup to estimate the range of plant growth strategies that lead to reproductive success under given climatic conditions. The simulated large-scale pattern from this modeling approach reproduced the observed features of plant species richness very well, with characteristic gradients in diversity along moisture and temperature gradients. The functional diversity of organisms as simulated by their approach can be interpreted here as degrees of freedom associated with the macroscopic process of biotic activity. These biotic degrees of freedom also introduce flexibility to the macroscopic processes of water and carbon cycling.

We can also understand the applicability of MEP to the biota in terms of macroscopic reproducibility (see also Dewar, 2003; Dewar, this volume; Lineweaver, this volume). When addressing the role of the biota at the macroscale, we do not require microscopic reproducibility, but are interested in the reproducibility of the macroscopic state. This may be illustrated as follows: Given certain values of annual mean precipitation and temperature, we can with high certainty predict whether this climate would likely lead to a tropical rainforest, a grassland, a tundra, or a desert. The vegetation type in turn determines the macroscopic state of the land surface, in terms of its surface albedo, its aerodynamic roughness, and its rooting zone depth. What we do not know is whether the community at the microscale at a certain location is composed of species A, B, and C at a given time t. But for the macroscopic description of the land surface and for the functional consequences for the global biogeochemical cycles of water and carbon, this information is not required, except to the extent to which sufficient biotic degrees of freedom are represented. Macroscopic reproducibility then tells us that the MEP state is the most likely macroscopic state, that is, the macroscopic MEP state can be reproduced by the vast majority of microstates which in our case would be the compositions of communities formed by individual plants.

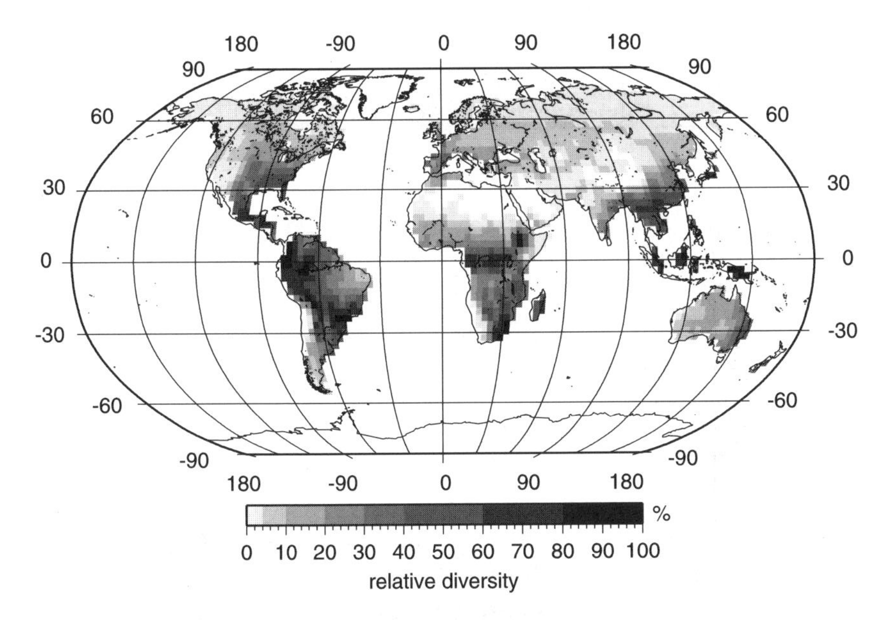

Fig. 14.1. Simulated plant functional diversity by a modeling approach of Kleidon and Mooney (2000). A simulation model of a generic plant was used in conjunction with a Monte-Carlo simulation to estimate the range of feasible plant growth strategies that lead to reproductive success under different climatic conditions. The geographic variation of diversity is interpreted here as biotic degrees of freedom, which are a necessary requirement for MEP to apply to biotic activity

14.4.2 Biotic States of MEP

We suggest two MEP states relevant to atmosphere-biosphere interactions at the large scale, which are both related to the Earth's albedo. As discussed in the beginning of this chapter, the Earth's planetary albedo is not fixed, but variable and flexible, determined primarily by the surface albedo and the extent of cloud and snow cover. The planetary albedo plays a crucial role of the overall rate of entropy production of planet Earth (14.1), and small changes in the planetary albedo can dwarf the contributions of other

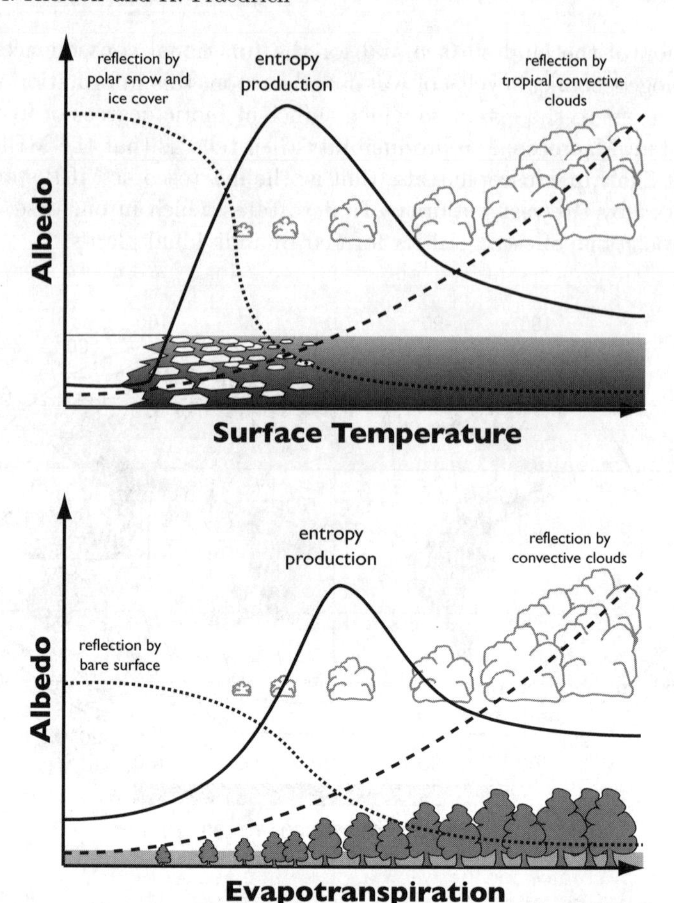

Fig. 14.2. Conceptual diagrams of how the (*top*) planetary albedo and (*bottom*) land surface albedo exhibit a minimum with respect to surface temperature and land surface evapotranspiration. A minimum in the overall albedo leads to a maximum in absorption of solar radiation and therefore to a maximum in entropy production (*solid lines*). Since surface temperature is related to the strength of the atmospheric greenhouse, which in turn is affected by the biota through its effects on biogeochemical cycles, there are potentially many possible states with a range of global mean temperatures that satisfy the constraints of global energy- and carbon balance. Likewise, the energy- and water budget constraints on land surface functioning permit potentially many states with differing rates of evapotranspiration, ranging from a bare surface to a fully vegetated one. The MEP principle in these two cases states that the state of MEP is the most likely macroscopic state of the system. See text for further explanations

processes to the overall entropy budget of Earth (see Kleidon and Lorenz, this volume). The two states of MEP can be explained as follows (Fig. 14.2):

- *biotic MEP and carbon cycling:* Biotic activity affects atmospheric concentrations of carbon dioxide (and methane), and therefore the strength of the atmospheric greenhouse effect and surface temperature. To illustrate the state of MEP, let us think of surface temperature as an external control parameter at the planetary scale for a moment, meaning that we have multiple potential steady states with different values of surface temperature that satisfy the constraints of the global energy and carbon budget. We would likely expect the following effects of surface temperature on planetary albedo: With decreasing surface temperature, the extent of snow cover would increase in the polar regions, leading to an increase in planetary albedo. On the other hand, increasing surface temperature leads to a lower equilibrium Bowen ratio (14.7), which leads to a moister boundary layer and less boundary layer growth. This should result in an increase in convective clouds, particularly in the tropics, and an associated increase in planetary albedo. Taken together, these two lines of reasoning suggest that there is a minimum planetary albedo at which the absorption of solar radiation is at a maximum. This, in turn, would lead to a state of maximum entropy production of the Earth system. Surface temperature of course is not an external control parameter, but at the planetary scale is determined largely by the absorption of solar radiation at the surface and the strength of the atmospheric greenhouse, which is affected by the extent of biotic carbon cycling and the biotic enhancement of rock weathering (see e.g., Schwartzman and Lineweaver, this volume). This leads then to a connection between a planetary state of MEP associated with absorption of solar radiation and biotic carbon exchange.
- *biotic MEP and water cycling:* The examples of a "Desert World" and a "Green Planet" suggest a state of MEP associated with evapotranspiration (which, as explained above, is directly linked to the productivity of terrestrial vegetation). Let us again view evapotranspiration as an external parameter for a moment that we can adjust, meaning that there are many macroscopic states of the land surface that satisfy the energy- and water balance. We get the following effects of the magnitude of evapotranspiration on net albedo: Increasing evapotranspiration allows for higher productivity, therefore allowing for a lower surface albedo to be maintained through more green biomass at the surface. This leads to increased absorption of solar radiation. On the other hand, increased evapotranspiration leads to a moister boundary layer with less boundary layer growth, resulting in increased formation of convective clouds. This leads to an increase of the net albedo of the atmospheric column and reduces the amount of incoming solar radiation at the surface. Therefore, there should be a state of MEP associated with evapotranspiration. This MEP state can be illustrated at a qualitative level by the extreme climate model sim-

ulations of a "Desert World", the "Present-Day", and a "Green Planet". Table 14.2 clearly shows the trend towards higher cloud cover with increasing presence of terrestrial vegetation as suggested in Fig. 14.2b (see also Kleidon 2004b). Note however that the vegetation state of a "Green Planet" does not account for productivity constraints on land surface parameters, that is, rainforest characteristics are prescribed even in desert climates, which could not be maintained in steady state due to lack of productivity. This lack of constraint may explain the absence of a peak in biotic productivity.

14.4.3 Biotic MEP and Gaia

If we accept biotic MEP as an emergent property of atmosphere-biosphere interactions in steady-state, then this can have important implications for the adaptability of climate system functioning to change. This is illustrated by the sensitivity of a simple coupled climate-carbon cycle model (Kleidon 2004a) to a prescribed external change in solar output of solar radiation (i.e., the solar luminosity, which affects I_0 in equation (14.1)). This model implements the line of reasoning described above for biotic MEP and carbon cycling (as shown in Fig. 14.2a). When the model is forced by changes in solar luminosity (which was 70% of today's value some 4 billion years ago) and it is assumed that biotic activity adjusts to a maximum in entropy production, then the resulting simulated surface temperature is insensitive to these changes (Fig. 14.3). The resulting evolution of atmospheric carbon dioxide concentrations associated with the MEP state is roughly similar to what reconstructions suggest about the past evolution of the atmospheric greenhouse (e.g., Kasting 1993; Catling, this volume). The homeostatic outcome of this simple model shares similarity to the Gaia hypothesis, which

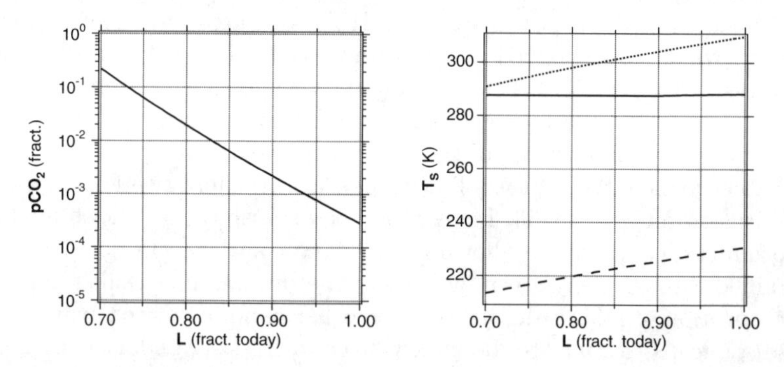

Fig. 14.3. Sensitivity of (*left*) atmospheric carbon dioxide concentrations pCO_2 and (*right*) surface temperature to variations of solar luminosity at a state of biotic MEP as simulated by a simple climate-carbon cycle model. Solar luminosity is expressed as a fraction of the present-day value. Also shown are the simulated temperatures for $pCO_2 = 1$ (*dotted line*) and $pCO_2 = 10^{-6}$ (*dashed line*). After Kleidon (2004a)

states "environmental homeostasis for and by the biosphere" (Lovelock and Margulis 1974; see also Toniazzo et al., this volume). The important difference is, however, that the emergent outcome of this example is the result of MEP as a physical selection principle, and biotic processes being viewed as representing additional degrees of freedom for climate system processes.

14.5 Conclusions

In this chapter we discussed the role of the biota for present-day climate system functioning, focusing on the role of terrestrial vegetation. We illustrated how climatic conditions constrain photosynthetic activity, and that these constraints are not fixed, but an emergent outcome of atmosphere-biosphere interactions and biotic feedbacks. For terrestrial vegetation, the overall feedbacks can be described by two positive feedbacks: the boreal forest feedback for temperature-limited regions and the water cycling feedback for water-limited regions. We showed that biotic entropy production at the large scale in steady-state can be approximated by the gross primary productivity, and that the two feedbacks are consistent with the notion that biotic effects enhance entropy production. The existence of a maximum in biotic entropy production was qualitatively demonstrated with two examples involving carbon exchange and evapotranspiration. Both examples specifically emphasize the role of convective clouds to understand biotic MEP states in the coupled atmosphere-biosphere system. We also illustrated that *if* we assume that the biosphere adjusts to MEP when environmental conditions (such as solar luminosity) change, then the outcome can lead to environmental homeostasis. This notion naturally needs to be substantiated, for instance with further modeling studies using process-based simulation models of the biosphere and the Earth system. Nevertheless, the perspective we promote here seems to be a promising path to appreciate the role of biodiversity in the functioning of the Earth system from a fundamental, thermodynamic perspective and to understand the ability of the Earth system to adapt to global changes.

Acknowledgements. We thank Lola Olsen for constructive comments on this manuscript. A. K. is partly supported by the General Research Board of the University of Maryland and the National Science Foundation through grant ATM-0336555.

References

Betts RA (1999) Self-beneficial effects of vegetation on climate in an ocean-atmosphere general circulation model. Geophys Res Lett 26: 1457–1460.

Bonan GB (2002) Ecological Climatology. Cambridge University Press, Cambridge, UK.

Bonan GB, Pollard D, Thompson SL (1992) Effects of boreal forest vegetation on global climate. Nature 359: 716–718.
Buckley TN, Farquhar GD, Mott KA (1999) Carbon-water balance and patchy stomatal conductance. Oecologia 118: 132–143.
Charney JG (1975) Dynamics of deserts and drought in the Sahel. Q J Roy Meteorol Soc 101: 193–202.
Claussen M (1994) On Coupling Global Biome Models with Climate Models. Clim Res 4: 203–221.
Claussen M (1998) On Multiple Solutions of the Atmosphere-Vegetation System in Present-Day Climate. Global Change Biol 4: 549–559.
Claussen M, Gayler V (1997) The greening of Sahara during the mid-Holocene: results of an interactive atmosphere – biome model. Global Ecol Biogeog 6: 369–377.
deNoblet NI, Prentice IC, Joussaume S, Texier D, Botta A, Haxeltine A (1996) Possible role of atmosphere-biosphere interactions in triggering the last glaciation. Geophys Res Lett 23: 3191–3194.
Dewar RC (2003) Information theory explanation of the fluctuation theorem, maximum entropy production, and self-organized criticality in non-equilibrium stationary states. J Physics A 36: 631–641.
Eltahir EAB (1998) A soil moisture-rainfall feedback mechanism. 1. Theory and observations. Water Resources Research 34: 765–776.
Field CB, Behrenfeld MH, Randerson JT, Falkowski P (1998) Primary production of the biosphere: integrating terrestrial and oceanic components. Science 281: 237–240.
Foley JA, Kutzbach JE, Coe MT, Levis S (1994) Feedbacks between climate and boreal forests during the Holocene epoch. Nature 371: 52–54.
Fraedrich K, Kleidon A, Lunkeit F (1999) A green planet versus a desert world: Estimating the effects of vegetation extremes on the atmosphere. J Clim 12: 3156–3163.
Gallimore RG, Kutzbach JE (1996) Role of orbitally induced changes in tundra area in the onset of glaciation. Nature 381: 503–505.
Hartmann DL (1994) Global physical climatology. Academic Press, San Diego CA.
Kasting JF (1993) Earth's early atmosphere. Science 259: 920–926.
Kleidon A (2002) Testing the effect of life on Earth's functioning: How Gaian is the Earth system? Clim Change 52: 383–389.
Kleidon A (2004a) Beyond Gaia: Thermodynamics of life and Earth system functioning. Clim Change, in press.
Kleidon A (2004b) Optimized stromatal conductance of vegetated land surfaces and its effects on simulated productivity and climate. Geophys Res Letters, in press.
Kleidon A, Fraedrich K, Heimann M (2000) A green planet versus a desert world: estimating the maximum effect of vegetation on land surface climate. Clim Change 44: 471–493.
Kleidon A, Heimann M (2000) Assessing the role of deep rooted vegetation in the climate system with model simulations: mechanism, comparison to observations and implications for Amazonian deforestation. Clim Dyn 16: 183–199.
Kleidon A, Lorenz S (2001) Deep roots sustain Amazonian rainforest in climate model simulations of the last ice age. Geophys Res Letters 28: 2425–2428.

Kleidon A, Mooney H (2000) A global distribution of biodiversity inferred from climatic constraints: results from a process-based modelling study. Glob Ch Biol 6: 507–523.

Kutzbach JE, Bonan G, Foley JA, Harrison SP (1996) Vegetation and soil feedbacks on the response of the African monsoon to orbital forcing in the early to middle Holocene. Nature 384: 623–626.

Larcher W (1995) Plant Physiological Ecology. 3rd edition. Springer Verlag, New York, NY.

Lovelock JE, Margulis L (1974) Atmospheric homeostasis by and for the biosphere: the Gaia hypothesis. Tellus 26: 2–10.

Milly PCD, Dunne KA (1994) Sensitivity of the global water cycle to the water-holding capacity of land. J Clim 7: 506–526.

Monsi M, Saeki T (1953) Über den Lichtfaktor in den Pflanzengesellschaften und seine Bedeutung für die Stoffproduktion. Jap J Bot 14: 22–52.

Monteith JL (1977) Climate and the efficiency of crop production in Britain. Phil Trans R Soc Lond B 281: 277–294.

Pitman AJ (2003) The evolution of, and revolution in, land surface schemes designed for climate models. Int J Climatol 23: 479–510.

Roeckner E, Arpe K, Bengtsson L, Christoph M, Claussen M, Dümenil L, Esch M, Giorgetta M, Schlese U, Schulzweida U (1996) The atmospheric General Circulation Model ECHAM-4: model description and simulation of present-day climate. Report 218, Max-Planck-Institut für Meteorologie, Hamburg, Germany. ISSN 0937-1060.

Schneider ED, Kay JJ (1994) Life as a manifestation of the second law of thermodynamics. Math Comput Modeling 19: 25–48.

Ulanowicz RE, Hannon BM(1987) Life and the production of entropy. Proc R Soc Lond B 232: 181–192.

15 Coupled Evolution of Earth's Atmosphere and Biosphere

David C. Catling

Astrobiology Program, Department of Atmospheric Sciences, University of Washington, Box 351640, Seattle, WA 98195, USA

Summary. Earth's climate has remained conducive to life for more than 3.5 billion years, probably because of higher concentrations of CO_2 and CH_4 greenhouse gases in the past that compensated for a much lower solar flux. Both CO_2 and CH_4 are involved in biogeochemical negative feedback loops that help to stabilize climate. The most significant event in Earth's atmospheric history was an increase in atmospheric O_2. The energetic advantage of aerobic metabolism facilitated larger levels of anabolism (the conversion of simple molecules to complex molecules), which gave rise to multicellular life, anatomical differentiation, and diverse biological complexity. The need for abundant O_2 to facilitate complex life on any planet is probably universal given the thermodynamic uniqueness of oxygen in the periodic table. Earth's atmosphere has evolved to a state that is chemically and physically distinctive compared to the atmospheres on the other planets. Chemically, Earth's atmosphere is continuously maintained at a low entropy state by the biosphere. Physically, Earth is the sunniest planet in terms of energy flux on its surface, while the atmosphere has the most unpredictable weather systems and, paradoxically, the slowest atmospheric jets of any planet with an atmosphere.

15.1 Introduction

The Earth – replete with oceans, forests, and animals – differs vastly from any other planet in the Solar System. However, this cannot have always been so and the question of how the complex world around us developed from lifeless beginnings provides a great interdisciplinary challenge. The geochemical record of ancient rocks and the chemical influence of the modern biosphere tell us that the evolution of life has effected major changes in the chemistry of the atmosphere and oceans. Essentially, the cycling of all elements of use to biology has been partially co-opted by the biosphere. The evolution of oxygenic photosynthesis and its effect on the cycling of elements arguably wrought the biggest change of all. Consequently, the history of atmospheric oxygen provides the environmental framework for discussing the evolution of biological complexity. Without free molecular oxygen, energetic limitations would have precluded life from ever becoming complex. On the other hand, the development of an oxygen-rich atmosphere was perhaps inevitable once bacteria evolved the capability to split water by oxygenic photosynthesis. Coupled with physical process of hydrogen escape to space from the atmosphere, the Earth's atmosphere was destined to become more oxidized at some point. Once oxygen exceeded a critical threshold, animal life and multicellular plants were the products of opportunistic evolution. Thus, understanding the details of the history of atmospheric oxygen is a key part of the puzzle in understanding how Earth has evolved to be the Solar System's chemical anomaly and how life on Earth evolved to states of lower entropy.

15.2 The Earliest Earth: Its Atmosphere and Biosphere

15.2.1 What Was the Composition of the Prebiotic Atmosphere?

An important question concerning the interaction of the earliest biosphere with the atmosphere is the extent to which the early atmosphere provided a mixture of chemically reducing and oxidized gases of potential metabolic use. Today, the consensus view is that before life appeared, the atmosphere was likely to have been "weakly reducing", composed primarily of N_2, CO_2 and water, with relatively small quantities of H_2, CO, and CH_4, and negligible O_2 (Walker 1977; Kasting and Brown 1998). N_2 was probably similar to present levels, having outgassed early (Fanale 1971). CO_2 may have been present at higher levels (although see Zahnle and Sleep (2000) for an alternative view). Laboratory experiments that produced organic molecules, such as amino acids, in CH_4-NH_3-H_2 atmospheres, had led to suggestions that the early atmosphere was highly reducing to explain how life originated (e.g., Miller 1955). This view has now been largely abandoned, in part, because Rubey (1955) introduced the notion that the atmosphere derived from gases emanating from the solid Earth. Rubey's belief that gradual degassing had led to a slowly increasing atmospheric mass has been replaced by a view of early, rapid degassing (Fanale 1971), but his insight about the redox state of gases entering the atmosphere still holds. Rubey noted that volatiles introduced to the atmosphere from modern degassing are only weakly reducing. Oxidized species such as H_2O, CO_2, and N_2 dominate over the reduced forms, H_2, CO or CH_4, and NH_3. The reduced to oxidized gas ratio (H_2/H_2O, CO_2/CO, etc.) in volcanic gases depends on the degree of oxidation in the upper mantle, the source region for such gases (Holland 1984). Geochemical evidence suggests that much of the mantle's oxidation was completed early, before 3.9 Ga (Canil 1997, 1999, 2002; Delano 2001), possibly during core formation before 4.4 Ga. Hence degassed volatiles have been weakly reducing for most of Earth history and the steady-state prebiotic atmosphere should have reflected this abiotic redox balance.

15.2.2 When Did Earth Acquire a Biosphere?

No one knows exactly when life originated but constraints can be deduced from the geological record and the expected physical environment of the early

Earth. Large impacts that pummeled the early Earth cannot be ignored. In particular, impact objects with size > 440 km would have vaporized all the oceans, sterilizing the planet. Calculations of the occurrence of the last sterilizing impact put a bound on the timing of the origin of the last common ancestor to no earlier than ∼ 4.3–3.8 Ga (Sleep et al. 1989). Geology can tell us when the earliest life is observed. Carbon globules from > 3.7 Ga marine metasedimentary rocks in west Greenland are enriched in ^{12}C by an amount consistent with biological fractionation (Rosing 1999). Reports of older, 3.8–3.9 Ga carbon isotope indicators of life (Mojzsis et al. 1996) are now doubted (Fedo and Whitehouse 2002; van Zuilen et al. 2002). While 3.5 Ga "microfossils" (Schopf 1993) are disputed (Brasier et al. 2002), 3.23–3.47 Ga spherical, carbonaceous microfossils from S. Africa are plausibly biogenic because some are in the process of binary division like cells (Knoll and Barghoorn 1977). Thread-like microfossils also occur in 3.2 Ga deep-sea volcanic rocks (Rasmussen 2000). That microbial life was globally present in the Archean Eon (before 2.5 Ga) can also be inferred from geochemistry. Marine carbonates back to 3.5 Ga are depleted in ^{12}C by an amount similar to modern marine limestone, suggesting that a global biosphere extracted the ^{12}C complement into organic matter (Schidlowski 1988). Indeed, the average organic carbon content of Archean marine shales (fine-grained sedimentary rocks) is ∼ 0.5 wt.% , essentially no different from that in recent geological eras. All of this suggests that a global microbial biosphere was present by 3.5 Ga (Nisbet and Sleep 2001).

15.2.3 What Effect Did Primitive Life Have on the Early Atmosphere?

Even the most primitive biosphere would have affected atmospheric composition, shifting it to a greater state of chemical disequilibrium. Today's microbial biosphere affects the cycling of every major element of importance to biology (C, N, O, P, S and so on) and early microbial life no doubt behaved similarly. Life modulates the cycle of carbon, the second most abundant volatile after water, by removing carbon from the atmosphere to synthesize organic matter. Life also modulates the cycle of Earth's third most abundant volatile, nitrogen, by extracting it from the air to make ammonium ions, and then recycling it back to the air. In fact, when early metabolisms evolved they must have particularly affected atmospheric H_2, CO_2, and N_2, because H_2 is food, CO_2 provides a carbon source, and nitrogen is essential for the peptide bond that holds organisms together.

Life would have caused atmospheric H_2 levels to drop (Fig. 15.1), with hydrogen atoms returning to the atmosphere in the form of less edible methane. Methanogens derive energy from hydrogen and carbon dioxide that would have co-existed in the prebiotic atmosphere derived from volatiles continuously degassed from the solid Earth:

$$4H_2 + CO_2 \rightarrow CH_4 + 2H_2O \qquad (15.1)$$

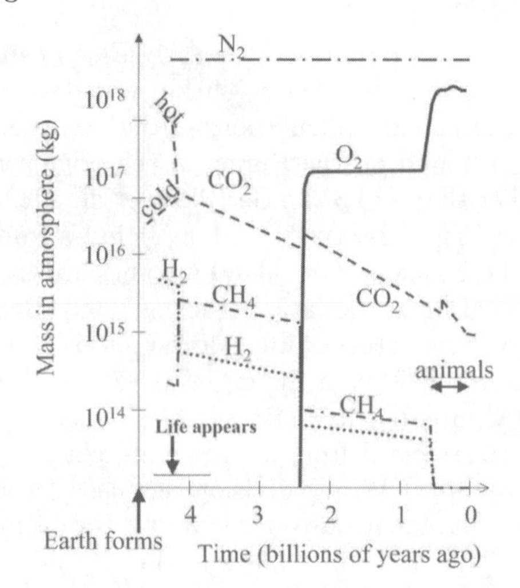

Fig. 15.1. A rough schematic history of the Earth's atmospheric composition. The bifurcation in possible early CO_2 levels reflects the fact that some make the case for a cold, icy start to the early Earth (Sleep and Zahnle 2001), while others argue for a hot beginning (see Schwartzman and Lineweaver, this volume)

The free energy is used by methanogens to synthesize cell material from inorganic nutrients. Assuming that methanogens were thermodynamically limited, $\geq$ 90% of the H_2 would have been converted to CH_4 (Kasting et al. 2001), consistent with laboratory experiments (Kral et al. 1998). Consequently, H_2 levels would have dropped from a prebiotic mixing ratio of 10^{-3} to about $\sim 10^{-4}$ once methanogenesis evolved. CO_2 levels would have also decreased because CH_4 is a powerful greenhouse gas, which would have led to increased warming and loss of CO_2 via temperature-dependent weathering.

Life must have also affected the nitrogen cycle. Nitrogen is often a limiting nutrient because few microorganisms can metabolize nitrogen unless it is fixed from the atmosphere and turned into a soluble form. The first organisms must have relied upon abiotic sources of soluble nitrogen. In the anoxic prebiotic atmosphere, N_2 would have been oxidized with CO_2 by lightning:

$$N_2 + 2CO_2 \rightarrow 2NO + 2CO \quad (15.2)$$

NO dissolves to produce NO_3^- and NO_2^- as end products (Mancinelli and McKay, 1988), providing a modest flux of fixed nitrogen (Navarro-Gonzales et al. 2001). Ammonium could have also been produced from HCN hydrolysis because HCN is synthesized in atmospheres containing trace levels of CH_4. In any case, once anaerobes developed nitrogen fixation, NH_4^+ would likely have become the dominant combined form of nitrogen in the ocean and this should have expanded the realm of the biosphere. Nitrogen fixation is strictly anaerobic (Postgate 1987; Zehr et al. 1995) and its origin may even predate the divergence of the three domains of life (Fani et al. 2000). After oxygenic photosynthesis arose (see below), nitrification must have become important. Nitrification is where ammonium ions are oxidized to nitrite and then nitrate by aerobic bacteria. Denitrification, the microbial reduction of NO_3^- using organic matter to N_2O and N_2, completes the cycle. Thus after the $\sim$ 2.3 Ga rise of O_2 (Fig. 15.1), the nitrogen cycle was probably similar to today's.

The overall effect of early life, then, was to shift the atmosphere away from a state of modest chemical disequilibrium maintained by sunlight to a state of much greater disequilibrium.

15.3 Long-Term Climate Evolution and the Biosphere

Modification of the Earth's atmospheric composition by the biosphere would have affected greenhouse warming. Earth is warmed by absorption of low entropy visible and near-infrared radiation from the Sun and is cooled by emission of high entropy thermal infrared radiation. In between, the Earth and biosphere do dissipative work. If we treat the Earth as a blackbody with effective temperature T_e, the equilibrium energy flux balance is

$$S/4\,(1-A) - \sigma T_e^4 = 0 \tag{15.3}$$

Here, $\sigma(= 5.67 \times 10^{-8}\,\mathrm{W/m^2/K^4})$ is the Stefan-Boltzmann constant, $S(= 1370\,\mathrm{W/m^2})$ is the solar flux at Earth's orbit, and A ($\cong 0.3$), is the planetary albedo, or reflectivity. Solving for the effective temperature yields $T_e = 255\,\mathrm{K}$. However, Earth is not a blackbody and the atmosphere warms the surface to temperature $T_s \sim 288\,\mathrm{K}$ through the greenhouse effect, which has a magnitude $\Delta T = T_e - T_s = 33\,\mathrm{K}$. Today H_2O vapor is responsible for about 2/3 of the greenhouse warming, while CO_2 accounts for much of the remainder. CH_4, N_2O, O_3, and anthropogenic chlorofluorocarbons contribute $\sim$ 2–3 K of the total.

The greenhouse effect should have been larger in the past (Sagan and Mullen 1972). Solar models indicate that the Sun was about 30% less bright when it formed 4.5 b.y. ago and that its luminosity has increased more or less linearly with time (Gough 1981). If one reduces S by 30% in (15.1), holding A and ΔT_g constant, T_e drops to 233 K and $T_s = 266\,\mathrm{K}$, well below the freezing point of water. Actually, the problem is more severe if the strong temperature dependence of the water vapor content of air is included. Radiative-convective models suggest that T_s drops below the freezing point of water prior to $\sim$ 2 Ga ago, if the bulk gases of the atmosphere were constant. However, geologic evidence tells us that liquid water and life were both present back to at least 3.5 Ga (Sect. 15.2). The solution to this problem likely resides in increased concentrations of greenhouse gases. Indeed, there are good reasons why both CO_2 and CH_4 should have been more abundant in the deep past.

For CO_2, the argument involves the carbonate-silicate cycle (Walker et al. 1981; Berner et al. 1983). If the surface temperature were cold, the rate

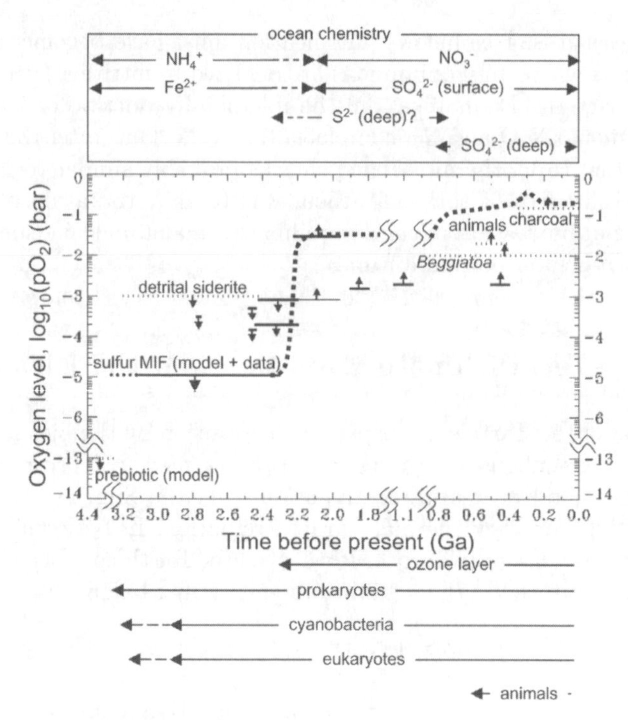

Fig. 15.2. The co-evolution of atmospheric O_2, ocean chemistry and life. The *thick dashed line* shows a possible evolutionary path for atmospheric O_2 that satisfies biogeochemical data. *Dotted horizontal lines* show the duration of biogeochemical constraints, such as the occurrence of detrital siderite ($FeCO_3$) in ancient riverbeds. Downward-pointing arrows indicate upper bounds on the partial pressure of oxygen (pO_2), whereas upward-pointing arrows indicate lower bounds. Unlabelled *solid horizontal lines* indicate the occurrence of particular paleosols (ancient soils), with the length of each line showing the uncertainty in the age of each paleosol. Inferences of pO_2 from paleosols are taken from Rye and Holland (1998). An upper bound on the level of pO_2 in the prebiotic atmosphere at c. 4.4 Ga (shortly after the Earth had differentiated into a core, mantle and crust) is based on photochemical calculations. MIF stands for "mass-independent isotope fractionation", which in sulfur is caused by photochemistry an O_2-poor atmosphere. The pO_2 level inferred from MIF observed in pre-2.4 Ga sulfur isotopes is based on the photochemical model results of Pavlov and Kasting (2002). Biological lower limits on pO_2 are based on the O_2 requirements of: (1) the marine sulfur-oxidizing bacterium, *Beggiatoa* (Canfield and Teske 1996); (2) animals that appear after 0.59 Ma (Runnegar 1991); (3) charcoal production in the geologic record. A "bump" in the oxygen curve around ~ 300 Ma, in the Carboniferous, is based on the interpretation of Phanerozoic carbon and sulfur isotope data by Berner et al. (2000)

of silicate weathering would be low because of the dependence of chemical reaction rates on temperature and also as a result of low rainfall. Since weathering ordinarily removes CO_2, volcanic supply would increase the CO_2 concentration in the atmosphere. Eventually some surface temperature would be reached that supports rainfall and a carbon cycle. This feedback stabilizes global temperature on a timescale of 10^5–10^6 years.

Although CO_2 may have been somewhat higher, geological evidence shows that Earth's early atmosphere had very little O_2 until 2.3 Ga, and this implies that another greenhouse gas, methane, should have been predominant. Today biogenic methane is rapidly destroyed by oxidation, with a net reaction of $CH_4 + 2O_2 = CO_2 + H_2O$. Oxidation limits atmospheric methane to 1.7 ppmv. However, in the absence of O_2, methane would be long-lived and considerably more abundant. Photochemical models (Zahnle 1986; Pavlov et al. 2001) show that the lifetime of CH_4 in a low-O_2 atmosphere is $\sim$5,000–10,000 years, as opposed to $\sim$10 years today. Consequently, the present biological methane flux of 535 Tg CH_4/yr (Houghton 1994) could have supported an atmospheric CH_4 mixing ratio of $\sim$300 ppmv (Pavlov et al. 2001). The methane flux would likely have been substantial because the prebiotic H_2 mixing ratio would be $\sim 10^{-3}$ at steady-state with a balance of H_2 outgassing and escape to space. So CH_4 mixing ratios after methanogenesis evolved should have been roughly half of this, given efficient biological conversion by reaction (15.1). If CH_4 was an abundant constituent of the Archean atmosphere, then the greenhouse effect could have been large even if CO_2 concentrations were relatively modest (Pavlov et al. 2000) as geochemical evidence indicates (Rye et al. 1995; see also discussion in Catling et al. 2001).

15.4 Atmospheric Redox Change: The Rise of Oxygen

Abundant CH_4 in the early atmosphere comes at a price. Methane is photolyzed in the upper atmosphere by ultraviolet (UV) light below 145 nm wavelength. As a result hydrogen is liberated and escapes to space. Hydrogen escape oxidizes the Earth as a whole. Earth's hydrogen was not captured gravitationally from the solar nebula, so it ultimately comes from condensed materials such as hydrated silicates, organic compounds, and water ice. When hydrogen is lost from such materials, the residue is oxidized. Abundant atmospheric methane in the Archean would promote rapid escape of hydrogen to space, oxidizing the Earth, and it also would counteract the fainter sun by greenhouse warming. Thus, the major Earth history problems of the "faint young Sun" and "rise of O_2" are linked through methane (Catling et al. 2001).

Methanogenesis is a more ancient metabolism than oxygenic photosynthesis on the basis of genetic studies and biochemical complexity (e.g., Hedges et al. 2001). Before oxygenic photosynthesis evolved, methane would have been derived from the decomposition of organic matter made by anoxygenic photosynthesizers or chemoautotrophs. Such prokaryotes use H_2, reduced sulfur, or Fe^{2+} as electron donors (e.g., $2H_2S + CO_2 + h\nu \rightarrow CH_2O + H_2O + 2S$). CH_4 derived from such organic matter (via $2CH_2O = CH_4 + CO_2$) and subsequent H escape to space would cause oxidized sulfur or iron to be gained irreversibly by the crust or mantle given that the electron donor ultimately originates from the crust or mantle. Free O_2 is not produced. After the advent of oxygenic photosynthesis, a much larger supply of organic matter would have been available, and the methane flux to the atmosphere should have increased substantially. Given that reducing chemicals scavenged oxygen out of Archean atmosphere, paradoxically the immediate effect of the advent of oxygenic photosynthesis was probably an increase in atmospheric CH_4, promoting further oxidation of the Earth through H escape. Le Châtelier's principle[1] demands that atmospheric O_2 sinks decrease as the crust is irreversibly oxidized via CH_4-induced H escape. Inevitably, at some point O_2 was destined to flood the atmosphere.

The atmosphere started out with virtually no oxygen ($pO_2 \sim 10^{-13}$ bar) before life existed and probably had only a few ppmv in the late Archean (Pavlov and Kasting 2002). The subsequent rise of atmospheric O_2 at ~2.4–2.2 Ga irrevocably changed the course of biological evolution (Fig. 15.2). At ~2.4–2.2 Ga, the onset of red beds, oxidized soils, a step change in mass-independent sulfur isotopes, and the oxidative loss of detrital reduced minerals in riverbeds all indicate an "oxic transition" in the atmosphere (Catling and Kasting 2005). The partial pressure of oxygen (pO_2) is estimated to have been < 0.0008 atm before the transition and possibly > 0.03 atm afterwards (Rye and Holland 1998). Paleontology is consistent with the hypothesis that the rise of O_2 triggered biological evolution. The oldest fossils visible to the naked eye are the remains of a spiral seaweed, *Grypania spiralis* (Han and Runnegar 1992) from shales now dated at 1.87 Ga. Acritarchs, which are fossilized organic tests of unicellular algae, are first found in 1.8 Ga rocks in China and become abundant in younger rocks. Fossils that are 4–5 mm long are found in northern China at 1.7 Ga, and may be multicellular (Shixing and Huineng 1995). Then by 1.4 Ga, similar, often larger, carbonaceous fossils become abundant worldwide in marine sedimentary rocks. A subsequent increase in pO_2 at 1.0–0.64 Ga is inferred from an increase in sulfate in the ocean (Canfield and Teske 1996; Canfield 1998). The appearance of multicellular animals in the fossil record at ~0.6 Ga must have depended upon sufficient atmospheric O_2 as a precursor.

[1] Le Châtelier's principle states that if a dynamic equilibrium is disturbed by changing conditions, the position of equilibrium moves to counteract the change. Atmospheric oxygen has a tendency to be removed by reaction with reducing gases released from the solid Earth (i.e., oxygen + reducing gases = oxidized gases). An irreversibly more oxidized crust and the consequent "stress" of the release of fewer reducing gases from the solid Earth would cause the atmospheric dynamic equilibrium to shift to the left according to Le Châtelier's Principle. This would produce higher oxygen levels.

Another important effect of the rise in O_2 at 2.4–2.2 Ga was to create a stratospheric ozone layer, shielding surface life from harmful UV radiation. Radiation below about 200 nm is strongly absorbed by CO_2 even in the absence of significant O_2, but ozone also shields biologically harmful radiation in the 200–300 nm range. Photochemical models show that harmful UV is mostly absorbed with an ozone layer that would form with O_2 levels about 1–3% of present (Kasting and Donahue 1980), similar to those inferred after 2.3 Ga (Fig. 15.2).

15.5 Oxygen, Energy, and Life

The increase in the amount of O_2 in the Earth's atmosphere gave organisms the potential for a much more energetic metabolism. Combined with evolution, this new energetic capability first produced larger organisms and ultimately organisms with differentiated anatomy, including brains and intelligence.

15.5.1 Aerobic Versus Anaerobic Energetics

Aerobic respiration provides about an order of magnitude more energy than anaerobic metabolisms. In living cells, energy is conserved first as adenosine triphosphate or ATP. ATP is used primarily in the synthesis of macromolecules and to drive active transport across cell membranes. The aerobic degradation of glucose, $C_6H_{12}O_6 + 6O_2 = 6CO_2 + 6H_2O$ (ΔG° – -2877 kJ/mol), yields 31 mol of ATP per mol glucose. The subsequent hydrolysis of ATP to provide energy has a free energy change of -29.3 kJ/mol. Thus, the efficiency of aerobic energy conversion is 32% ((29.3 kJ/mol $\times$31 mol)/ 2877 kJ)), with the other 68% lost as metabolic heat. In contrast, fermentative heterotrophic bacteria yield only $\sim$2–4 mol ATP per mol glucose. For example, the anaerobe *Lactobacillus* that turns milk sour ferments glucose to lactic acid ($C_6H_{12}O_6 = 2CH_3CHOHCOOH$, $\Delta G^\circ = -197$ kJ/mol) yielding 2 mol of ATP. This implies a 30% efficiency ((29.3 kJ/mol $\times$2 mol)/ 197 kJ)), similar to aerobic metabolism, but produces only 1/16 of the energy of aerobic metabolism.

The significant lack of free energy in anaerobic metabolism precludes the capability for multicellular growth and multiple trophic levels. A useful generalization is that the growth yield per mol ATP is $\sim$ 10 g dry weight of organic matter (written as $Y_{ATP} = 10$ g/mol in bacteriology) (Bauchop and Elsden 1960; Russell and Cook 1995; Desvaux et al. 2001). For aerobic metabolism, 310 g ($31 \times Y_{ATP}$) of growth can be produced by ingesting 500 g of glucose (180 g/mol glucose + 310 g), which is a 63% (310/490) growth efficiency. In contrast, an average anaerobic metabolism producing 3 mol ATP per mol glucose, generates only 30 g of growth for an intake of 210 g, which is only a 14% growth efficiency. These numbers indicate the general relative growth efficiency between aerobic and anaerobic lifestyles, noting that measurements on higher animals show quite a range of 30–50% in growth efficiency (Brafield and Llewellyn 1982). The metabolic restriction on growth efficiency means that only aerobic organisms grow large. Aerobic respiration is efficient enough to allow eukaryotes to eat other organisms and grow large and multicellular in a hierarchical food chain. But anaerobic organisms are doomed to a unicellular lifestyle.

The evolution of metazoans required sufficient atmospheric O_2 not just for respiration but also for the synthesis of many essential biochemicals (Margulis et al. 1976). The latter includes sterols (e.g., cholesterol), polyunsaturated fatty acids that comprise membranes, the amino acids tyrosine and hydroxyproline, collagen (a hydroxyproline-containing structural protein) (Towe 1981), and biochemicals used in sclerotisation to harden cuticles (Brunet 1967). The first metazoans must have relied on O_2 diffusion. In conditions of low O_2($\sim$ 1% of present levels), these creatures were necessarily small ($<$ 0.5 mm) because diffusion is proportional to the O_2 gradient from the exterior (Raff and Raff 1970; Runnegar 1982). Circulatory systems with respiratory proteins and mineral skeletons only evolved once O_2 levels were higher (Runnegar 1991). Metazoans emerged in a late burst of rapid diversification $\sim$560–540 Ma ago (Ayala et al. 1998). This time is after a second increase in O_2 inferred from the geochemical record (Knoll 1992; Knoll and Holland 1995; Canfield and Teske 1996).

15.5.2 Why Complex Life Anywhere in the Universe Will Likely Use Oxygen

Free, diatomic oxygen (O_2) is clearly necessary for the growth and survival of terrestrial complex life. But is substantial free O_2 a universal requirement for complex life or is it merely peculiar to the vagaries of evolution on Earth? We can pose this question in another way: Given the need for a copious supply of energy by complex life, what is the most energetic chemical reaction available? Obviously, more energy is released if there are more reactants, so comparisons of energy release must be normalized per electron transfer. For any combination of elements in the periodic table, H-F and H-OH bonds have the highest bond enthalpies per electron (Table 15.1), giving the most energetic reactions per electron transfer. Table 15.1 compares terminal oxidants ranked according to the Gibb's free energy (ΔG^0) for reduction to an aqueous hydride. Because ΔG^0 ($= \Delta H^0 - T\Delta S^0$) is dominated by enthalpy changes (ΔH^0), it generally tracks the hydride bond enthalpy. In the gas phase, oxygen is the next most energetic oxidant after fluorine, although in the aqueous phase (as shown), chlorine is marginally more energetic than oxygen. An extraterrestrial metabolism using fluorine would provide the most energy but for the fact that fluorine explodes in contact with organic matter.

Table 15.1. Comparison of possible terminal oxidants for biology. The list is ranked according to free energy of reduction to the aqueous hydride per electron under standard conditions, ΔG° (298.15 K, 1 atm)

Atom	Solar System abundance (% atoms)[a]	Redox reaction with hydrogen equivalent	ΔG^0 (kJ)[b]	Hydride bond enthalpy[c] per electron (kJ mol^{-1})	Reaction products with water	Prohibitive kinetic reactivity with organics?
F	2.7×10^{-6}	$\frac{1}{2} F_2 + \frac{1}{2} H_2 = HF(aq)$	−296.82	−282.5	O_2, HF	Yes
Cl	1.7×10^{-5}	$\frac{1}{2} Cl_2 + \frac{1}{2} H_2 = HCl(aq)$	−131.23	−215.5	HOCl	Yes
O	0.078	$\frac{1}{2} O_2 + \frac{1}{2} H_2 = \frac{1}{2} H_2O(l)$	−118.59	−231.5	-	No
Br	3.8×10^{-8}	$\frac{1}{2} Br_2 + \frac{1}{2} H_2 = HBr(aq)$	−103.96	−183.0	HOBr	Yes
N	0.01	$\frac{1}{6} N_2 + \frac{1}{2} H_2 + \frac{1}{3} H^+ = NH_4^+(aq)$	−26.44	−194.0	-	No
S	0.0017	$\frac{1}{2} S + \frac{1}{2} H_2 = \frac{1}{2} H_2S(aq)$	−13.92	−169.0	-	No
C	0.033	$\frac{1}{2} C + \frac{1}{2} H_2 = \frac{1}{4} CH_4(aq)$	−8.62	−206.0	-	No

a. Anders and Grevesse (1989).
b. Lide (1997).
c. Pauling (1960).

Fluorine also reacts with every other element apart from a few of the noble gases. Similarly, in water, chlorine and bromine form reactive sterilants: hypochlorous acid (HOCl) and hypobromous acid (HOBr), respectively. O_2 does not react with water but merely dissolves because its double bond provides greater stability than the single bonds in the halogens. In any case, a halogen-rich planetary atmosphere is simply ruled out because gases like fluorine or chorine would be removed extremely rapidly by gas-solid, gas phase or aqueous reactions. Given that F_2 and Cl_2 could never be abundant gases in a planetary atmosphere or plausible biological oxidants, one must conclude that the presence of free O_2 in a planetary atmosphere allows life to utilize the most efficient energy source per electron transfer available within the periodic table. This is undoubtedly a universal property because the periodic table and thermodynamics are universal. There are also two other reasons why complex life would use oxygen. First, O is third in cosmological abundance behind H and He (Anders and Grevesse 1989). Second, oxygen is found in water. Liquid water is required for Earth-like life (Kushner 1981) and its presence on a planet sets the conventional definition for planetary habitability (e.g., Catling and Kasting 2003). If a planet is habitable, with liquid H_2O and life, it has biological potential for accumulating O_2 in its atmosphere, since this comes from biological splitting of H_2O in photosynthesis. In turn, such a planet would also have the potential for complex life.

15.6 The Anomalous Nature of Earth's Current Atmosphere

Lovelock (1965) noted that the low entropy chemical state of Earth's atmosphere was a sign of life on the Earth's surface, cause by large gas fluxes from the biosphere. In essence, the atmosphere is an extension of the low entropy biosphere. For example, given $pO_2 \sim 0.209$ bar, the thermodynamic equilibrium value of pCH_4 should be 10^{-145} bar. This means that there should not be a single molecule of CH_4 in Earth's oxygenated atmosphere at thermodynamic equilibrium. Instead, pCH_4 is $10^{-5.76}$ bar. As a life detection mechanism, however, there are potential ambiguities. Even an abiotic world would have chemical disequilibrium because free energy is supplied by solar energy, geothermal heat from radioactive decay, and tidal heating. Thus, chemical disequilibrium in a planetary atmosphere as a sign of life is a question of degree. The application of Maximum Entropy Production (MEP) to atmospheric chemistry may help us understand the distinction between biotic and abiotic planets.

I close this chapter with the further observation that Earth's atmosphere is physically anomalous. Whether life plays a part in this physical oddity (perhaps by affecting cloud or planetary albedo, for example, see Kleidon and Fraedrich, this volume) remains to be elucidated. The anomalous nature of Earth's atmosphere is a combination of the following three factors: Of planets with atmospheres, Earth is the sunniest in the Solar System (in terms of the flux on its surface) with the most unpredictable large-scale weather and the slowest atmospheric jets. Large-scale weather on other planets tends to be much more predictable. For example, the Red Spot of Jupiter is obviously highly predictable, having persisted for over 200 years. Similarly, Martian large-scale weather is sufficiently predictable that a hurricane-like weather system imaged by the Hubble Space Telescope was regenerated almost exactly about a year later, as imaged by the Mars Global Surveyor camera. Earth's unpredictable weather requires much more information for its description. Vern Suomi, the father of satellite meteorology first noted the jet stream paradox (Conway Leovy, personal communication), but it has not received any discussion in the literature. Suomi noticed that on planets from Earth to Neptune, the characteristic speed of the jets varies inversely with the available upward energy flux that drives the dynamics (Table 15.2). It is probably also important to realize that rates of kinetic energy dissipation are clearly low in the outer planet atmospheres but high in Earth's troposphere. Because of this, intuitively, it seems likely that this paradox will be explained through consideration of the principle of MEP.

Table 15.2. Suomi's paradox: The characteristic jet speed in planetary atmospheres *increases* despite a *decrease* of thermal forcing going from Earth to Neptune. (Courtesy of Conway Leovy)

Planet	Thermal Forcing [normalized to Earth]	Characteristic Jet Speed [m/s]
Earth	1.0	30
Mars	0.09	80
Jupiter	0.05	100
Saturn	0.02	150
Uranus, Neptune	0.003	300

References

Anders E, Grevesse N (1989) Abundances of the elements – Meteoritic and solar. Geochim. Cosmochim. Acta 53:197–214.

Ayala FJ, Rzhetsky A, Ayala FJ (1998) Origin of the metazoan phyla: Molecular clocks confirm paleontological estimates. Proc. Nat. Acad. Sci. 95:606–611.

Bauchop T, Elsden SR (1960) The growth of micro-organisms in relation to their energy supply. J. Gen . Microbiol. 23:457–469.

Berner RA, Lasaga AC, Garrels RM (1983) The carbonate-silicate geochemical cycle and its effect on atmospheric carbon dioxide over the past 100 million years. Amer. J. Sci. 283:641–683.

Berner R, Petsch ST, Lake JA, Beerling DJ, Popp BN, Lane RS, plus six authors. (2000) Isotope fractionation and atmospheric oxygen: Implications for phanerozoic O_2 evolution. Science 287: 1630–1633.

Brafield AE, Llewellyn MJ (1982) Animal Energetics, Chapman and Hall, New York.

Brasier MD, Green OR, Jephcoat AP, Kleppe AK, Van Kranendonk MJ, Lindsay JF, Steele A, Grassineau NV (2002) Questioning the evidence for the Earth's oldest fossils. Nature 416: 76–81.

Brunet PCJ (1967) Sclerotins. Endeavour 26:68–74.

Canfield DE (1998) A new model for Proterozoic ocean chemistry. Nature 396:450–453.

Canfield DE, Teske A (1996) Late-Proterozoic rise in atmospheric oxygen concentration inferred from phylogenetic and sulfur isotope studies. Nature 382:127–132.

Canil, D (1997) Vanadium partitioning and the oxidation state of Archaean komatiite magmas, Nature 389: 842–845.

Canil, D (1999) Vanadium partitioning between orthopyroxene, spinel and silicate melt and the redox states of mantle sources regions for primary magmas. Geochim. Cosmo. Acta 63: 557–572.

Canil, D (2002) Vanadium in peridotites, mantle redox and tectonic environments: Archean to present. Earth Planet. Sci. Lett. 195: 75–90.

Catling, D. and Kasting, J. F. (2005) Planetary atmospheres and life. In Sullivan, W. T. and Baross, J. (eds) Planets and Life: The Emerging Science of Astrobiology. Cambridge, Cambridge University Press. In press.

Catling DC, Zahnle KJ, McKay CP (2001) Biogenic methane, hydrogen escape, and the irreversible oxidation of early Earth. Science 293:839–843.

Delano, JW (2001) Redox history of the Earth's interior: implications for the origin of life. Origins Life Evol. Biosphere 31: 311–341.

Desvaux M, Guedon E, Petitdemange H (2001) Kinetics and metabolism of cellulose cegradation at high substrate concentrations in steady-state continuous cultures of *Clostridium Cellulolyticum* on a chemically defined medium. Appl. Environ. Microbiol. 67:3837–3845.

Fanale FP (1971) A case for catastrophic early degassing of the Earth. Chem. Geol. 8:79–105.

Fani R, Gallo R, Lio P (2000) Molecular evolution of nitrogen fixation: the evolutionary history of the *nif*D, *nif*K, *nif*E, and *nifN* genes. J Mol. Evol 51:1–11.

Fedo CM, Whitehouse MJ (2002) Metasomatic origin of quartz-pyroxene rock, Akilia, Greenland, and implications for Earth's earliest life. Science 296:1448–1452.

Gough DO (1981) Solar interior structure and luminosity variations. Solar Phys. 74:21–34.

Han T-M, Runnegar B (1992) Megascopic eukaryotic algae from the 2.10billion-year-old Negaunee iron-formation, Michigan. Science 257:232–235.

Hedges SB, Chen H, Kumar S, Wang DYC, Thompson AS, Watanabe H (2001) A genomic timescale for the evolution of eukaryotes. BMC Evol. Biology 1:1–10.

Holland HD (1984) The Chemical Evolution of the Atmosphere and Oceans, edn. Princeton University Press, Princeton.

Houghton JT, et al. (1994) Climate Change, 1994: Radiative Forcing of Climate Change and an Evaluation of the IPCC IS92 Emission Scenarios, Cambridge University Press, Cambridge.

Kasting JF, Brown LL (1998) Setting the stage: the early atmosphere as a source of biogenic compounds. In: Brack, A. (ed) The Molecular Origins of Life: Assembling the Pieces of the Puzzle. New York, Cambridge University Press. pp. 35–56.

Kasting JF, Donahue TM (1980) The evolution of atmospheric ozone. J. Geophys. Res. 85: 3255–3263.

Kasting JF, Pavlov AA, Siefert JL (2001) A coupled ecosystem-climate model for predicting the methane concentration in the Archean atmosphere. Origins of Life Evol. Biosph. 31:271–285.

Kasting JF, Catling D (2003) Evolution of a habitable planet. Ann Rev Astron Astrophys 41:429–463.

Knoll AH (1992) The early evolution of the eukaryotes: a geological perspective. Science 256:622–627.

Knoll AH, Barghoorn ES (1977) Archean microfossils showing cell division from the Swaziland System of South Africa. Science 198:396–398.

Knoll, AH. and Holland, HD (1995) Oxygen and Proterozoic evolution: An update. In: Effects of Past Global Change on Life. Washington, D.C., Nat. Acad. Press.

Kral TA, Brink KM, Miller SL, McKay CP (1998) Hydrogen consumption by methanogens on the early Earth. Origins of Life and Evol. of the Biosph. 28:311–319.

Kushner D (1981) Extreme environments: Are there any limits to life? In: Ponnaperuma C (ed) Comets and the Origin of Life. Reidel, Dordrecht, pp. 241–248.

Lide DR (ed) (1997) Handbook of Chemistry and Physics, CRC Press, Boca Raton, FL.

Lovelock JE (1965) A physical basis for life detection experiments. Nature 207:568–570.

Mancinelli RL, McKay CP (1988) The evolution of nitrogen cycling. Origins of Life 18:311–325.

Margulis L, Walker JCG, Rambler MB (1976) Reassessment of roles of oxygen and ultraviolet light in Precambrian evolution. Nature 264:620–624.

Miller SL (1955) Production of some organic compounds under possible primitive Earth conditions. J. Am. Chem. Soc. 77:2351–2361.

Mojzsis SJ, Arrhenius G, McKeegan KD, Harrison TM, Nutman AP, Friend CRL (1996) Evidence for life on Earth before 3,800 million years ago. Nature 384:55–59.

Navarro-Gonzalez, R, McKay, CP, Mvondo, DN (2001) A possible nitrogen crisis for Archean life due to reduced nitrogen fixation by lightning. Nature 412: 61–64.

Nisbet EG, Sleep NH (2001) The habitat and nature of early life. Nature 409:1083–1091.

Pauling L (1960) The nature of the chemical bond. Cornell Univ. Press.

Pavlov AA, Kasting JF (2002) Mass-independent fractionation of sulfur isotopes in Archean sediments: strong evidence for an anoxic Archean atmosphere. Astrobiology 2:27–41.

Pavlov AA, Kasting JF, Brown LL (2001) UV-shielding of NH_3 and O_2 by organic hazes in the Archean atmosphere. J. Geophys. Res. 106:23,267–23,287.

Pavlov AA, Kasting JF, Brown LL, Rages KA, Freedman R (2000) Greenhouse warming by CH_4 in the atmosphere of early Earth. J. Geophys. Res. 105:11,981–11,990.

Postgate, JR (ed) (1987) Nitrogen fixation. Edward Arnold, London.

Raff RA, Raff EC (1970) Respiratory mechanisms and the metazoan fossil record. Nature 228:1003–1005.

Rasmussen B (2000) Filamentous microfossils in a 3,235-million-year-old volcanogenic massive sulpgide deposit. Nature 405:676–679.

Rosing MT (1999) 13C-depleted carbon microparticles in $>$ 3700-Ma sea-floor sedimentary rocks from West Greenland. Science 283:674–676.

Rubey, WW (1955) Development of the hydrosphere and atmosphere, with special reference to probably composition of the early atmosphere. In: Crust of the Earth. A. Polervaart (ed) Geological Society of America, New York, pp. 631–650.

Runnegar B (1982) Oxygen requirements, biology and phylogenetic significance of the late Precambrian worm Dickinsonia, and the evolution of the burrowing habit. Alcheringa 6:223–239.

Runnegar, B (1991) Oxygen and the early evolution of the Metazoa. In Bryant, C (ed) Metazoan Life without Oxygen. Chapman and Hall, New York.

Russell JB, Cook GM (1995) Energetics of bacterial-growth: Balance of anabolic and catabolic reactions. Microbiol. Rev. 59:48–62.

Rye R, Kuo PH, Holland HD (1995) Atmospheric carbon dioxide concentrations before 2.2 billion years ago. Nature 378:603–605.
Rye R, Holland HD (1998) Paleosols and the evolution of atmospheric oxygen: A critical review. Amer. J. Sci. 298: 621–672.
Sagan C, Mullen G (1972) Earth and Mars: Evolution of atmospheres and surface temperatures. Science 177:52–56.
Schidlowski M (1988) A 3,800-million-year isotopic record of life from carbon in sedimentary rocks. Nature 333:313–318.
Schopf JW (1993) Microfossils of the Early Archean Apex chert: new evidence for the antiquity of life. Science 260:640–646.
Shixing Z, Huineng C (1995) Macroscopic cellular organisms from 1,700 million year old Tuanshanzi Formation in the Jixian Area, North China. Science 270:620–622.
Sleep NH, Zahnle KJ, Kasting JF, Morowitz HJ (1989) Annihilation of ecosystems by large asteroid impacts on the early Earth. Nature 342:139–142.
Sleep NH, Zahnle KJ (2001) Carbon dioxide cycling and implications for climate on ancient Earth. J Geophys Res 106:1373–1399.
Towe, K. M. (1981) Biochemical keys to the emergence of complex life. In: Billingham, J. Life in the Universe. MIT Press, Cambridge, MA, pp. 297–306.
van Zuilen MA, Lepland A, Arrhenius G (2002) Reassessing the evidence for the earliest traces of life. Nature 418:627–630.
Walker JCG (1977) Evolution of the Atmosphere, Macmillan, New York.
Walker JCG, Hays PB, Kasting JF (1981) A negative feedback mechanism for the long-term stabilization of Earth's surface temperature. J. Geophys. Res. 86:9776–9782.
Zahnle KJ (1986) Photochemistry of methane and the formation of hydrocyanic acid (HCN) in the Earth's early atmosphere. J. Geophys. Res. 91:2819–2834.
Zahnle KJ, Sleep NH (2002) Carbon dioxide cycling through the mantle and implications for the climate of ancient Earth, In: Fowler CMR, Ebinger CJ, Hawkesworth CJ (eds) The Early Earth: Physical, Chemical and Biological Development. Geological Society, London. Special Publications 199, pp. 231–257.
Zehr JP, Mellon M, Braun S, Litaker W, Steppe T, Paerl HW (1995) Diversity of heterotrophic nitrogen-fixation genes in a marine cyanobacterial mat. Appl. Environ. Microbiol. 61:2527–2532.

16 Temperature, Biogenesis, and Biospheric Self-Organization

David Schwartzman[1] and Charles H. Lineweaver[2]

[1] Department of Biology, Howard University, Washington, DC 20059, USA
[2] Department of Astrophysics and Optics, University of New South Wales, Sydney, 2052, Australia

Summary. We argue that the biosphere has evolved deterministically as a self-organized system, given the initial conditions of the Sun-Earth system. With temperature as a critical constraint, biogenesis and the overall patterns of biotic evolution were the highly probable outcomes of this deterministic process. Emergence of life and its major evolutionary innovations occurred as soon as the temperature decreased to their upper temperature limits. These innovations include phototrophy, oxygenic photosynthesis, and the emergence of Eukarya ("complex" life) and its Kingdoms.

16.1 Introduction

Tim Lenton (1998) raised a fundamental question: How can self-regulation at all levels of the biosphere emerge from natural selection at the individual level? A seemingly unrelated but deeply connected debate centers around a challenge to a long held view of evolutionary biology, namely that if the "tape" of life's history were played again, the results would be radically different owing to the stochastic nature of evolutionary emergence. The alternative is that the evolution of the biosphere is deterministic, i.e., its history and the general pattern of biotic evolution is very probable, given the same initial conditions. Likewise the evolution of self-regulating biospheres on Earth-like planets around Sun-like stars is deterministic. We would reverse Tim Lenton's question: was the pattern of natural selection determined by the self-regulating history of the biosphere? As Conway Morris (1998–1999) put it, "[biotic] history is constrained, and not all things are possible".

More precisely, we argue that the evolution of life and the biosphere is quasi-deterministic, i.e., the general pattern of the tightly coupled evolution of biota and climate was very probable and self-selected from a relatively small number of possible histories at the macroscale, given the same initial conditions (see also discussion on macroscopic reproducibility in Lineweaver, this volume). Major events in biotic evolution were likely forced by environmental physics and chemistry, including photosynthesis as well as the merging of complementary metabolisms that resulted in new types of cells (such as eucaryotes) and multicellularity. Determinism likely breaks down at finer levels. Critical constraints on this deterministic aspect of biotic evolution have likely included surface temperature, along with oxygen and carbon dioxide levels (see Catling, this volume).

Why quasi-deterministic? In addition to events that were probably inevitable, it is likely there is a role for randomness in both abiotic and biotic evolution even on the coarsest scale. For example, this randomness may include the influence of large impacts on the history of life, the possible multiple attractor states in mantle convection and therefore plate tectonic history, even multiple attractors for steady-state climatic regimes. Thus, there are a finite number of histories, for the same initial conditions and cosmic background, hence "quasi"- deterministic evolution.

Evolution of procaryotes and complex life on terrestrial planets around Sun-like stars are expected to have similar geochemical and climatic consequences. Thus, the main patterns would be conserved if "the tape were played twice", a theory argued from computer simulations by Fontana and Buss (1994). The width of the habitable zone for Earth-like planets around Sun-like stars for complex life may be substantially smaller than that for the appearance of biota, constrained only by the presence of liquid water. Surface temperature history on terrestrial planets may be critical to the time needed to evolve complex life and intelligence. For Earth-like planets within the habitable zone of stars less massive than the Sun, the earlier emergence of complex life is expected, all other factors being the same, since its upper temperature limit is reached earlier as a result of the lower rate of luminosity increase (Schwartzman 1999).

16.1.1 Cosmology and Temperature

Big Bang cosmology has given us an abiotic, deterministic model for the evolution of the Universe in which, as the Universe expanded and cooled from arbitrarily high temperature, an increasingly complex series of structures emerged including life and biospheres at least on terrestrial planets around Sun-like stars (see Lineweaver and Schwartzman, 2004; Chaisson, this volume; Lineweaver, this volume). A deterministic origin of life is now a virtual astrobiological paradigm, whether the preferred scenario invokes the primordial broth, hydrothermal regime or some variation with or without a significant extraterrestrial organic supply.

Here we examine the hypothesis that temperature has not only played the dominant deterministic role in cosmology and planet formation but also in biogenesis and biological evolution. That the diversification of life is strongly constrained by temperature and that as the temperature on the surface of the Earth decreased, this allowed the first life to appear but also determined what kind of life could appear.

16.2 Biogenesis at Life's Upper Temperature Limit: A Hyperthermophilic Origin of Life

The case for a high temperature origin of life was made by Wachtershauser (1998), who argued that biogenesis and microbial evolution proceeded deterministically from hyperthermophiles to mesophiles. We have likewise argued

(Schwartzman 1999; Lineweaver and Schwartzman 2004) that biogenesis and the emergence of the three domains of life was likely thermally inevitable, given the abiotic initial conditions of our planet. The chemical evolution research program continues to generate support for determinism, for example, in arguments for the likely universality of intermediary autotrophic metabolism (Morowitz et al. 2000) and the genetic code (Vogel 1998).

A hyperthermophilic last common ancestor of life (LCA) has long been thought implied by the rRNA phylogenetic tree (Fig. 16.1) (Woese 1987; Pace 1997). While the deep-rootedness of hyperthermophilic Archaea still appears to be robust (Matte-Tailliez et al. 2002; Caetano-Anolles 2002), some researchers argue that even deeply – rooted Bacteria acquired their hyperthermophily by horizontal gene transfer (*hgt*) from Archaea (Aravind et al. 1998; Forterre et al. 2000; Brochier and Philippe 2002), a view under challenge (Di Giulio 2003a,b,c). All known hyperthermophiles apparently have reverse gyrase (*rg*) in a fused gene (Forterre 2002), with the exception of the newly discovered Nanoarchaeota (Waters et al. 2003). The latter discovery suggests that fusion of separate genes to reverse gyrase occurred at hyperthermophilic, not moderately thermophilic or mesophilic conditions as previously argued (Lopez-Garcia 1999; Declais et al. 2000). Alternatively, another protective mechanism was present in a hyperthermophilic LCA and early organisms (Baross 1998; Musgrave et al. 2002).

An objection to a hyperthermophilic LCA based on G + C content of rRNA of extant organisms (Galtier et al. 1999) has been challenged (Di Giulio 2000a,b 2003). In addition to G + C content, Di Giulio used a thermophily index based on the propensity of amino acids to enter more frequently into (hyper)thermophile proteins, concluding that the late stage of genetic code structuring took place in a (hyper)thermophilic organism. These studies have used the correlation of the optimal growth temperature with the above measures to infer the temperature of the LCA, but the maximum temperature for growth (T_{MAX}) is plausibly closer to the temperature of emergence, giving a stronger inference of a hyperthermophilic LCA. Similarly, a hyperthermophilic LCA is inferred from the robust correlation of T_{MAX} with the rRNA phylogenetic distance from the LCA for (hyper)thermophilic procaryotes (Fig. 16.2) (Schwartzman and Lineweaver 2004), with an inferred temperature exceeding 120° C, consistent with the newly discovered record T_{MAX} of 121° C for life (an Archaea close to Pyrodictium; Kashefi and Lovley 2003), although this reported T_{MAX} is now being challenged (Stetter 2003).

Other recent support for a hyperthermophilic LCA comes from expanded sequence and secondary structure data of rRNA (Shepard et al. 2003), the tRNA sequence tree (Tong et al. 2003) and experimental tests using modified enzyme residues (Miyazaki et al. 2001; Yamagishi et al. 2003). If future phylogenetic trees based on greater sampling and better understood genomes confirm the apparent near universal absence of hyperthermophiles with mesophilic ancestors (Fig. 16.1), the asymmetric evolution of hyperthermophiles to mesophiles would be confirmed. This asymmetry is supported for example, by the absence of any living eucaryote with a T_{MAX} exceeding a few degrees above 60° C, in spite of at least 2 billion years of opportunity to adapt to hyperthermophily.

A hyperthermophilic LCA does not of course require an origin of life at similar high temperatures. In particular, biogenesis could have occurred at mesophilic temperatures to be followed by a major impact event sterilizing the surface and leaving a subsurface hyperthermophilic LCA survivor or primitive procaryotes close to the LCA (Gogarten-Boekels et al. 1995), although a recent reinterpretation of the lunar impact record argues that such sterilizing events may well have been absent (Ryder 2003). Nevertheless, the possibility that biogenesis did indeed occur at hyperthermophilic temperatures should be reexamined in light of recent research. A plausible scenario for the origin of life requiring the presence of a hydrothermal regime and thus a proximate hyperthermophilic environment is outlined by Martin and Russell (2003), with the locus being FeS compartments produced by reaction of contrasting pH solutions reacting at the seafloor. They postulate a temperature for this origin at about 50° C, argued from the assumed constraint of RNA instability at higher temperatures (Moulton et al. 2000). However saline solutions, readily available in this environment, appear to stabilize DNA/RNA at hyperthermophilic temperatures (Marguet and Forterre 1998; Tehi et al. 2002). Therefore, the possibility of an RNA/DNA world (Dworkin et al. 2003) that is hyperthermophilic should be revisited. Other objections to a hot origin of life including the instability of amino acids and ribose at hyperthermophilic temperatures (Levy and Miller 1998; Islas et al. 2003) have been addressed by supporters of a hydrothermal scenario, citing evidence for the thermodynamic favorability (Shock et al. 1998) and actual synthesis of amino acids under hydrothermal conditions (Hennet et al. 1992; Marshall 1994).

Thus, we argue that the chemical evolution to the LCA of Bacteria and Archaea could have occurred at hyperthermophilic conditions in Martin and Russell's scenario as Russell himself argued previously (Russell et al. 1998). Hence, the divergent chemical evolution leading to Archaeal and Bacterial free-living cells as proposed by Martin and Russell (2003) can be accommodated in a temperature gradient between the hydrothermal source and ambient climatic temperature, with the upper temperature limits of Archaea (greater or equal to about 120° C) and Bacteria (at 95° C if not higher) corresponding to the first possible self-organization of each cell type with cell membranes and walls. In this scenario, biogenesis occurred with the emergence of protocells with minimal metabolism and replication at the "edge of stability" to use the phrase in Musgrave et al. (2002). The differences between Archaeal and Bacterial membrane lipid biosynthesis and cell wall biochemistry is central to Martin and Russell's hypothesis. Their thermal stability might be clues to the temperature regimes prior to the emergence of free-living cells.

The deep nesting of primitive metabolisms inferred from extant hyperthermophiles (Ronimus and Morgan 2002; Berry 2002) can be readily explained by this scenario. The emergence of Archaea has been linked to the metabolism of methanogenesis (Koch 1994; Kral et al. 1998), a plausible

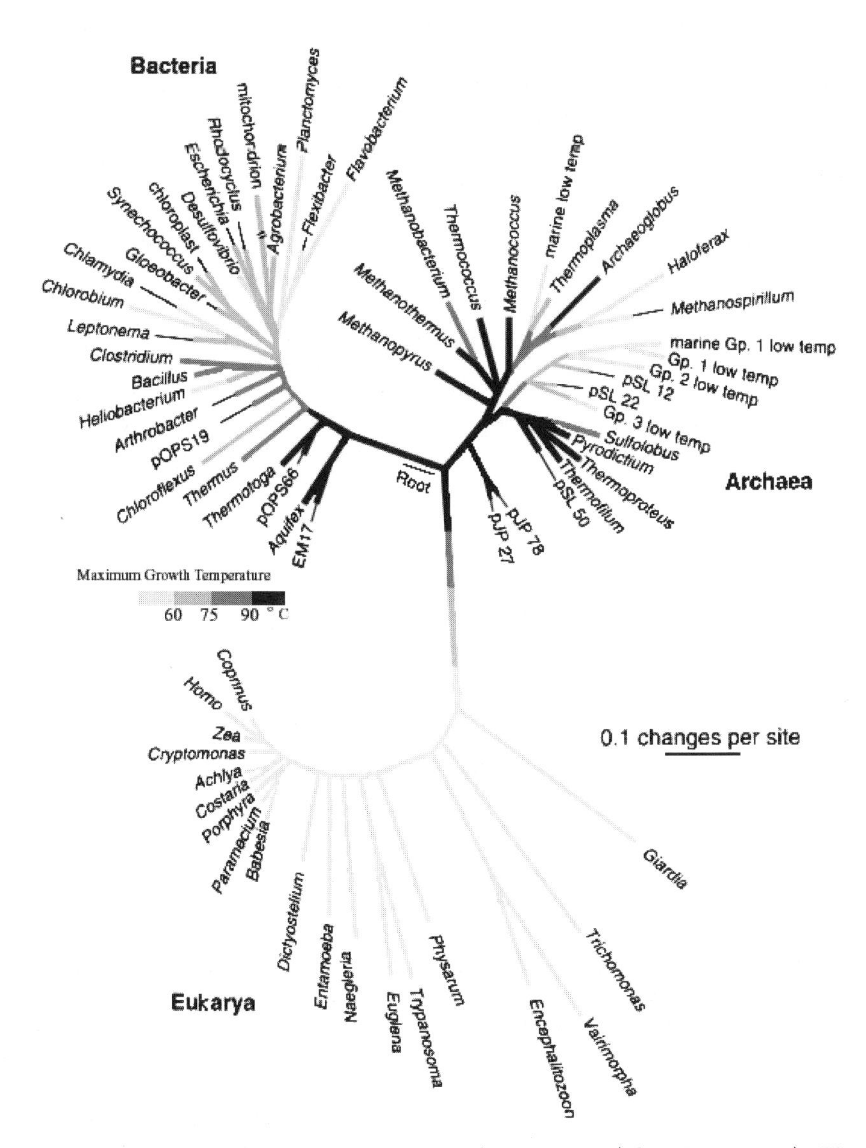

Fig. 16.1. Phylogenetic Tree based on rRNA sequences (after Pace, 1997). The scale bar corresponds to 0.1 changes per nucleotide. The last common ancestor (LCA) is located on the bacterial line between the node to Aquifex and the Y intersection of the three domains (Barns et al., 1996). Maximal growth temperatures (T_{MAX}) have been used to assign shades of grey to the branches (Lineweaver and Schwartzman, 2004). A thin black ending line indicates T_{MAX} data is lacking for this organism

natural exploitation of abundant carbon dioxide and hydrogen by early life (see Catling, this volume). The other primitive metabolisms found in deeply rooted Archaea can likewise be explained as the exploitation of available raw materials (sulfur, trace metals such as Fe, Ni, Cr, W and As; see Lebrun et al. 2003) readily available on early Earth, particularly in the hydrothermal regime environment (see Russell et al. 1998), first by chemoautotrophs soon followed by heterotrophs, in the chemoautotropic scenario for biogenesis. Plausibly the origin of these metabolisms occurred before 3.7 Ga, the age of metasedimentary rocks from Greenland containing the earliest known evidence for life, ^{13}C depleted carbon (Rosing 1999).

While the climatic temperature for the Hadean is not well constrained other than an inference of liquid water at 4.4 Ga (Valley et al. 2002), temperatures of 80° C at 3.8 Ga (Knauth 1992) and 70 (+/- 15)° C at 3.5 Ga (Knauth and Lowe 2003) have been derived from the oxygen isotopic record of marine cherts. Thus, while biogenesis plausibly occurred above late Hadean/early Archean climatic temperatures, a climatic temperature constraint on microbial evolution prevailed through the Precambrian with a temperature constraint on microbial evolution (Schwartzman 1999). The robust correlation of T_{MAX} with distance of the rRNA phylogenetic distance from the LCA (Figs. 16.1 and 16.2) is consistent with the maximum growth temperature of thermophiles, if not hyperthermophiles, being close to the climatic temperature of each organism at emergence.

16.3 The Temperature Constraint on Biologic Evolution

A temperature constraint on biologic evolution was first apparently proposed by Hoyle (1972) who suggested that a warm early Earth held back the emergence of low-temperature life. We argue that climatic (surface) temperature was the determining constraint with respect to the timing of major events in microbial evolution (e.g., emergence of photosynthesis, eucaryotes and Metazoa/Fungi/Plants), with thermophily prevailing for the first two-thirds of biospheric history. The biosphere evolved from a "hothouse" at its birth some 4 Ga ago to an "icehouse" for the last billion years, producing a biotic evolutionary explosion filling the myriad of new habitats created in the process.

However, surface temperature on Earth is not a fixed boundary condition for life, but it is affected by the biota. Surface temperature is regulated on a geological time scale by the carbonate-silicate biogeochemical cycle (Berner 1999; Schwartzman 1999). While on short time scales of less than 10^4 years the cycling between the atmosphere/ocean and surface pools such as organic carbon can have significant impact on atmospheric carbon dioxide levels (witness the glacial/interglacial cycles of the last 2 million years, anthropogenic impacts etc.), the long-term cycle on time scales greater than 10^5 years is controlled by the silicate-carbonate geochemical cycle. This cycle entails transfers of carbon to and from the crust and mantle. In the modern era, this cycle was first described by Urey in 1952:

$$CO_2 + CaSiO_3 = CaCO_3 + SiO_2 \quad (16.1)$$

The reaction to the right corresponds to chemical weathering of Ca silicates on land ($CaSiO_3$ is a simplified proxy for the diversity of rock-forming CaMg silicates such as plagioclase and pyroxene which have more complicated formulas, e.g., Ca plagioclase: $CaAl_2Si_2O_8$), while the reaction to the left corresponds to metamorphism ("decarbonation") and degassing returning carbon dioxide to the atmosphere.

This cycle is really biogeochemical. While decarbonation and outgassing is surely abiotic, taking place at volcanoes associated with subduction zones and oceanic ridges, chemical weathering involves biotic mediation, which entails soil stabilization by biotic cover on land along with the multifold processes induced by biological activity in the soil itself. The critical role of biology entails the progressive increase in the biotic enhancement of chemical weathering, the critical sink of carbon dioxide with respect to the atmosphere involving deposition of calcium carbonate on the ocean floor. This biotic enhancement of weathering generates a weathering flux balancing the volcanic source resulting in a steady-state level of atmospheric carbon dioxide (and temperature) at a lower level than on an abiotic Earth (see full discussion in Schwartzman 1999). On a global scale, the stabilization of thick soils by vascular plants makes possible the multifold accelerating effects of the rhizosphere, resulting in the estimated present biotic enhancement of weathering on the order of 10–100 times the abiotic regime. This biotically-mediated cooling increases the width of the habitable zone for the possible occurrence and evolutionary time frame of complex life.

If surface temperature was the critical constraint on microbial evolution, then the approximate upper temperature limit for viable growth of a microbial group should equal the actual surface temperature at time of emergence, assuming an ancient and necessary biochemical character determines the presently determined upper temperature limit T_{MAX} of each group (see Table 16.1, Fig. 16.3). The latter assumption is supported by an extensive data base of living thermophilic organisms. No phototroph has been found to grow above about 70° C, in spite of a likely ≥ 3.5 Ga age for this metabolism, similarly for eucaryotes with an upper limit of just over 60° C, with at least 2 Ga for the possibility of adaptation to life at higher temperatures. (Note: The report of the Pompeii worm living in hydrothermal vents with its attachment reaching 80° C is not good evidence that the upper temperature limit for metazoan growth exceeds 60 or even 50° C. Note that their head emerges from the vent chimney at 22° C (Cary et al. 1998)).

The upper temperature limit for viable growth is apparently determined by the thermolability of biomolecules (e.g., nucleic acids), organellar membranes and enzyme systems. For example, the mitochondrial membrane is particularly thermolabile, apparently resulting in an upper temperature limit of about 60° C for aerobic eucaryotes (note that the presumed ancestor of the mitochondrion, the Proteobacteria, have an upper temperature limit of 60° C).

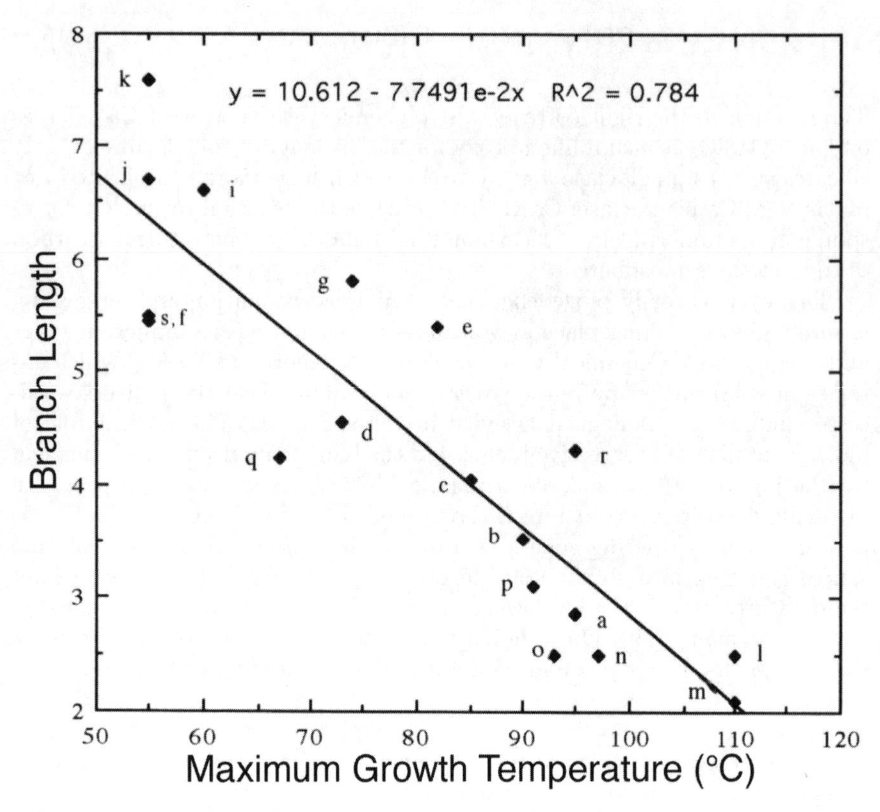

Fig. 16.2. The maximum growth temperature of (Hyper)Thermophiles versus their Phylogenetic Distance (Branch Length) from the LCA using rRNA phylogenetic tree (Fig. 16.1). The correlation coefficient r^2 for Archaea and Bacteria is $r^2 = 0.755$ and $r^2 = 0.777$ respectively. a Aquifex; b Thermotoga; c Thermus; d Chloroflexus; e Bacillus; f Chlorobium; g Synechococcus; i Agrobacterium; j Plantomyctes; k Flavobacterium; l Pyrodictium; m Methanopyrus; n Methanothermus; o Thermococcus; p Methanococcus; q Thermoplasma; r Archaeoglobus; s Haloferax For further details see Schwartzman and Lineweaver (2003)

Now let us return to the contention that the actual surface temperatures were thermophilic (greater than 50°C) for the first two-thirds of the history of the biosphere, i.e., from 4 Ga to about 1.5 Ga. This high temperature scenario for the Earth's climatic history remains controversial (see extensive discussion in Schwartzman 1999). The strongest evidence for this temperature scenario comes from the inferred paleotemperatures from the oxygen isotopic record of ancient chert and carbonates. Paleotemperatures have been derived from the oxygen isotope record of pristine cherts (Knauth and Lowe 1978; Knauth 1992; Knauth and Lowe 2003). The fractionation of the stable isotopes of oxygen is a function of the phases involved and temperature. A necessary assumption for these temperature calculations, that the Precambrian oxygen isotopic composition of seawater was close to the present, received recent support from studies of ancient seawater-altered oceanic crust. Robust new evidence for a very warm Archean comes from Knauth and Lowe's (2003) detailed study of the Onverwacht cherts, South Africa, centering on their oxygen isotopic record in the context of their geologic characteristics. The inferred paleotemperatures of ancient seawater are consistent with the inferred Precambrian temperature history illustrated in Fig. 16.3.

Table 16.1. Upper temperature limits for growth of living organisms and approximate times of their emergence. Temperatures from Brock et al. 1994

Group	Approximate upper temperature limit	Time of Emergence
	(°C)	(Ga)
Plants	50	$0.5^{\wedge}$–1.5^{*}
Metazoa	50	$0.6^{\wedge}$–1.5^{**}
Fungi	60	$0.4^{\wedge}$–1.5^{*}
Eucaryotes	60	2.1^{**}–$2.8^{\wedge\wedge}$
Procaryotic microbes		
Phototrophy	70	$\geq 3.5^{\wedge\#}$
Hyperthermophiles	≥ 80	$\geq 3.7^{\#}$

^ Fossil evidence
* Molecular phylogeny
** Problematic fossil evidence, molecular phylogeny
\# Biogenic carbon isotopic signature
^^ Organic geochemistry

We presume eucaryotes emerged as the endosymbiogenic product of Archaeal and Bacterial cells associating because of complementary metabolisms, with the acquisition of the mitochondrion close in time with the initial emergence of the eucaryotic cell (Martin and Russell 2003). Others have argued for a later acquisition of the mitochondrion (see e.g., Gupta 1998; Hedges et al. 2001) or even a primordial ancestor (the "Chronocyte") of the eucaryotic cell (Hartman and Fedorov 2002).

If the Chronocyte and amitochondrial eucaryotes arose earlier than the mitochondrial, then higher upper temperature limits are expected for both, with the Chronocyte perhaps in the hyperthermophilic range. In the two-step scenario for the origin of the mitochondrial eucaryotic cell, gram-negative bacteria are postulated to be fused with an archaeon, followed by the incorporation of the mitochondrion from a proteobacterium (Lang et al. 1999). If the T_{MAX} of at least 80°C of presently cultured gram-negative bacteria (e.g., Bacillus) is primitive then the emergence of amitochondrial eucaryotes may reach back into the early Archean.

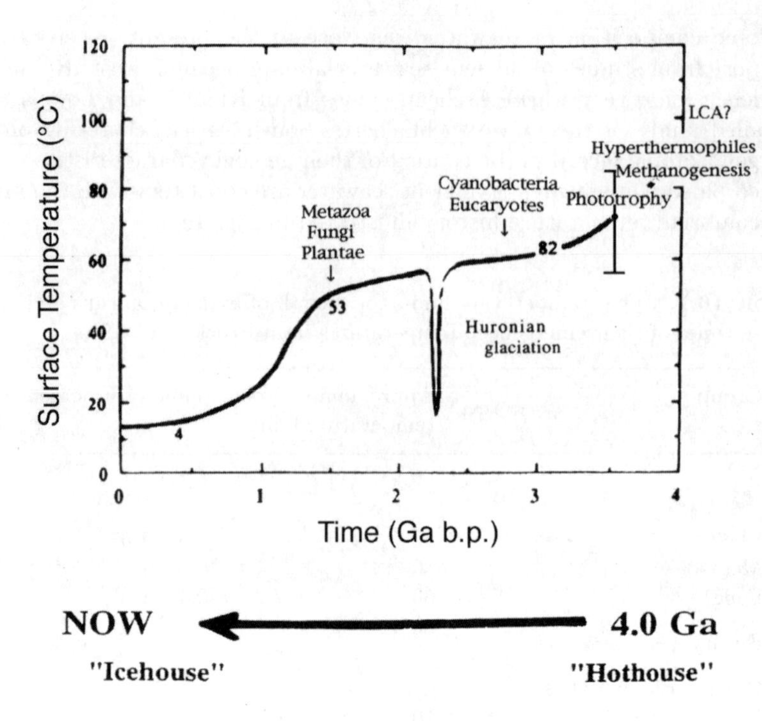

Fig. 16.3. Temperature of the surface of the Earth as a function of time. Emergence times of major organismal groups and long-term trends in biospheric evolution are shown. Numbers near curve are the computed model ratios of the present biotic enhancement of weathering to the value at that time (Schwartzman, 1999). The temperature for 3.5 Ga is from Knauth and Lowe (2003)

A deterministic emergence of the so-called higher kingdoms (Fungi, Metazoa and Plantae) is consistent with Szathmary and Maynard Smith's (1995) contention that epigenetic inheritance had already existed in protists prior to the emergence of these kingdoms. The inferred emergence time from molecular phylogeny (Heckman et al. 2001; Wang et al. 1999) and problematic metazoan trace fossils (e.g., Rasmussen et al. 2002a,b) are consistent with our temperature curve (Fig. 16.3).

The biogeochemical carbon cycle is linked to the oxygen cycle. Space precludes a full discussion of atmospheric levels of oxygen and carbon dioxide as constraints on evolution (see Catling, this volume). Aerobic microenvironments in the Archean apparently preceded the emergence of aerobic eucaryotes, which in turn preceded their first likely presence in the fossil record. Aerobic respiration may have emerged very early (Castresana and Saraste 1995). The rise of atmospheric oxygen by 1.9 Ga (15% present atmospheric level (PAL) according to Holland (1994) predated the emergence of Metazoa, which may require less than ~2% PAL (e.g., mud-dwelling nematodes; see Runnegar 1991). On the other hand, the atmospheric oxygen level plausibly constrained the emergence of megascopic eucaryotes, particularly Metazoa, as originally argued by Cloud (1976), with the explanation being the diffusion barrier of larger organisms (Raff 1996). The rise of a methane-dominated greenhouse and concomitant drop in atmospheric carbon dioxide level at about 2.8 Ga has been proposed as the trigger for the emergence of cyanobacteria (Schwartzman and Caldeira 2002).

16.4 Future Directions

Tests for (quasi-)deterministic evolution include further confirmation of the apparent close correspondence of organismal emergence times to the times that necessary environmental conditions are reached as inferred from the fossil, organic geochemical and paleoclimatic record, future discovery and characterization of alien biospheres (e.g., detection of atmospheric methane, oxygen and water) and computer simulations of biospheric evolution.

References

Aravind L, Tatusov RL, Wolf YI, Walker DR, Koonin EV (1998) Evidence for massive gene exchange between archaeal and bacterial hyperthermophiles. Trends in Genetics 14: 442–444.

Barns, SM, Delwiche CF, Palmer JD, Pace NR (1996) Perspectives on archaeal diversity, thermophily and monophyly from environmental rRNA sequences. Proc Natl Acad Sci USA 93: 9188–9193.

Baross JA (1998) Do the geological and geochemical records of the early earth support the prediction from global phylogenetic models of a thermophilic cenancestor? In: Wiegel J, Adams M (eds) Thermophiles: the keys to Molecular Evolution and the Origin of Life? Taylor and Francis, London, pp 3–18.

Berner RA (1999) A new look at the long-term carbon cycle. GSA Today 9: 1–6.

Berry S (2002) The chemical basis of membrane bioenergetics. J Mol Evol 54: 595–613.

Brochier C, Philippe H (2002) A non-hyperthermophilic ancestor for Bacteria. Nature 417: 244.

Brock TD, Madigan MT, Martinko JM, Parker J (1994) Biology of Microorganisms. 7th edition. Prentice Hall. Englewood Cliffs, NJ.

Caetano-Anolles G (2002) Evolved RNA secondary structure and the rooting of the universal tree of life. J Mol Evol 54: 333–345.

Castresana J, Saraste M (1995) Evolution of energetic metabolism: the respiration-early hypothesis. Trends in Biochemical Sciences 20: 443–448.

Cary SC, Shank T, Stein J (1998) Worms bask in extreme temperatures. Nature 391: 545–546.

Cloud P (1976) Beginnings of biospheric evolution and their biogeochemical consequences. Paleobiology 2: 351–387.

Conway Morris S (1998–1999) Showdown on the Burgess Shale. The Challenge. Natural History 107(10): 48–51.

Declais AS, Marsault J, Confalolnieri F, de la Tour CB, Duguet M (2000) Reverse gyrase, the two domains intimately co-operate to promote positive supercoiling. J Biol Chem 275: 19498–19504.

Di Giulio M (2000a) The universal ancestor lived in a thermophilic or hyperthermophilic environment. J Theor Biol 203: 203–213.

Di Giulio M (2000b) The late stage of genetic code structuring took place at a high temperature. Gene 261: 189–195.

Di Giulio M (2003a) The universal ancestor was a thermophile or a hyperthermophile: tests and further evidence. J Theor Biol 221: 425–436.

Di Giulio M (2003b) The ancestor of the Bacteria domain was a hyperthermophile. J Theor Biol 224: 277–283.

Di Giulio M (2003c) The universal ancestor and the ancestor of Bacteria were hyperthermophiles. J Mol Evol 57: 721–730.

Dworkin JP, Lazcano A, Miller SL (2003) The roads to and from the RNA world. J Theor Biol 222: 127–134.

Fontana W, Buss LW (1994) What would be conserved if "the tape were played twice"? Proc Nat Acad Sci USA 91: 757–761.

Forterre P (2002) A hot story from comparative genomics: reverse gyrase is the only hyperthermophile-specific protein. Trends in Genetics 18: 236–237.

Forterre P, de la Tour CB, Philippe H, Duguet M (2000) Reverse gyrase from hyperthermophiles: probable transfer of a thermoadaptation trait from Archaea to Bacteria. Trends in Genetics 16: 152–154.

Galtier N, Tourasse N, Gouy M (1999) A nonhyperthermophilic common ancestor to extant life forms. Science 283: 220–221.

Gogarten-Boekels M, Hilario E, Gogarten JP (1995) The effects of heavy meteorite bombardment on the early evolution – the emergence of the three domains of life. Orig Life Evol Biosph 25: 251–264.

Gupta RS (1998) Protein phylogenies and signature sequences: a reappraisal of evolutionary relationships among Archaebacteria, Eubacteria, and Eukaryotes. Microbiol. and Mol Bio Rev 62: 1435–1491.

Hartman H, Fedorov A (2002) The origin of the eukaryotic cell: a genomic investigation. Proc Natl Acad Sci USA 99(3): 1420–1425.

Heckman DS, Geiser DM, Eidell BR, Stauffer RL, Kardos NL Hedges SB (2001) Molecular evidence for the early colonization of land by fungi and plants. Science 293: 1129–1133.

Hedges SB, Chen H, Kumar S, Wang D Y-C, Thompson AS, Watanabe H (2001) A genomic timescale for the origin of eukaryotes. BMC Evol Biol 1: 4.

Hennet RJ-C, Holm NG, Engel, MH (1992) Abiotic synthesis of amino acids under hydrothermal conditions and the origin of life: a perpetual phenomenon? Naturwissenschaften 79: 361–365.

Hoyle F (1972) The history of the Earth. Quart J Roy Astron Soc 13: 328–345.

Holland HD (1994) Early Proterozoic atmospheric change. In: Bengtson S (ed) Early Life on Earth. Nobel Symposium No.84. Columbia University Press. New York, pp 237–244.
Islas S, Velasco AM, Becerra A, Delaye L, Lazcano A (2003) Hyperthermophily and the origin and earliest evolution of life. Int Microbiol 6: 87–94.
Kashefi K, Lovley DR (2003) Extending the upper temperature limit for Life. Science 301:934.
Knauth LP (1992) Origin and diagenesis of cherts: an isotopic perspective. In: Clauer N, Chaudhuri S (eds) Isotopic Signatures and Sedimentary Records, Lecture Notes in Earth Sciences, No 43 Springer-Verlag, Heidelberg, New York, pp 123–152.
Knauth LP, Lowe DR (1978) Oxygen isotope geochemistry of cherts from the Onverwacht Group (3.4 billion years), Transvaal, South Africa, with implications for secular variations in the isotopic composition of cherts. Earth Planet Sci Lett 41: 209–222.
Knauth LP, Lowe DR (2003) High Archean climatic temperature inferred from oxygen isotope geochemistry of cherts in the 3.5 Ga Swaziland Supergroup, South Africa. Geol Soc Am Bull 115(5): 566–580.
Koch AL (1994) Development and diversification of the last universal ancestor. J Theor Biol 168: 269–280.
Kral TA, Brink KM, Miller, SL, McKay CP (1998) Hydrogen consumption by methanogens on the early Earth. Orig Life Evol Biosph 28: 311–319.
Lang BF, Gray MW, Burger G (1999) Mitochondrial genome evolution and the origin of eukaryotes. Ann Rev Genetics 33: 351–397.
Lebrun E, Brugna M, Baymann F, Muller D, Lievremont D, Lett M-C, Nitschke W (2003) Arsenite oxidase, an ancient bioenergetic enzyme. Mol Biol Evol 20: 686–693.
Lenton TM (1998) Gaia and natural selection. Nature 394: 439–447.
Levy M, Miller, SL (1998) The stability of the RNA bases: implications for the origin of life. Proc Natl Acad Sci USA 95: 7933–7938.
Lineweaver CH, Schwartzman D (2004) Cosmic Thermobiology, thermal constraints on the origin and evolution of life in the universe. In: Seckbach J (ed) Origins: Genesis, evolution and biodiversity of microbial life in the Universe. Kluwer, pp. 233–248. astro-ph/0305214.
Lopez-Garcia P (1999) DNA supercoiling and temperature adaptation: a clue to early diversification of life? J Mol Evol 49: 439–452.
Marguet E, Forterre P (1998) Protection of DNA by salts against thermodegradation at temperatures typical for hyperthermophiles. Extremophiles 2: 115–122.
Marshall WL (1994) Hydrothermal synthesis of amino acids. Geochim Cosmochim Acta 58: 2099–2106.
Martin W, Russell MJ (2003) On the origin of cells: a hypothesis for the evolutionary transitions from abiotic chemistry to chemoautotrophic prokaryotes, and from prokaryotes to nucleated cells. Phil Trans R Soc London B 358: 59–85.
Matte-Tailliez O, Brochier C, Forterre P, Philippe H (2002) Archaeal phylogeny based on ribosomal proteins. Mol Biol Evol 19: 631–639.
Miyazaki J, Nakaya S, Suzuki T, Tamakoshi M, Oshima T, Yamagishi A (2001) Ancestral residues stabilizing 3-isopropylmalate dehydrogenase of an extreme thermophile: experimental evidence supporting the thermophilic common ancestor hypothesis. J Biochem 129: 777–782.

Morowitz HJ, Kostelnik JD, Yang J, Cody GD (2000) The origin of intermediary metabolism. Proc Natl Acad Sci USA 97(14): 7704–7708.

Moulton V, Gardner PP, Pointon RF, Creamer LK, Jameson GB, Penny D (2000) RNA folding argues against a hot-start origin of life. J Mol Evol 51: 416–421.

Musgrave D, Zhang X, Dinger M (2002) Archaeal genome organization and stress responses: implications for the origin and evolution of cellular life. Astrobiology 2: 241–253.

Pace NR (1997) A molecular view of microbial diversity and the biosphere. Science 276: 734–740.

Raff RA (1996) The Shape of Life. University of Chicago Press.

Rasmussen B, Bengtson S, Fletcher IR, McNaughton NJ (2002) Discoidal impressions and trace-like fossils more than 1200 million years old. Science 296: 1112–1115.

Rasmussen B, Bose PK, Sarkar S, Banerjee S, Fletcher IR, McNaughton NJ (2002) 1.6 Ga U-Pb zircon age for the Chorhat Sandstone, lower Vindhyan, India: possible implications for early evolution of animals. Geology 30: 103–106.

Ronimus RS, Morgan HW (2002) Distribution and phylogenies of enzymes of the Embden-Meyerhof-Parnas pathway from archaea and hyperthermophilic bacteria support a gluconeogenic origin of metabolism. Archaea 1: 1–23.

Rosing MT (1999) 13C-depleted carbon microparticles in $>$ 3700-Ma sea-floor sedimentary rocks from West Greenland. Science 283: 674–676.

Runnegar B (1991) Precambrian oxygen levels estimated from the biochemistry and physiology of early eukaryotes. Palaeogeogr Palaeoclimatol Palaeoecol 97: 97–111.

Russell MJ, Daia DE, Hall AJ (1998) The emergence of life from FeS bubbles at alkaline hot springs in an acid ocean. In: Wiegel J, Adams M (eds) Thermophiles: the keys to Molecular Evolution and the Origin of Life? Taylor and Francis, London, pp 77–126.

Ryder G (2003) Bombardment of the Hadean Earth: wholesome or deleterious? Astrobiology 3: 3–6.

Schwartzman D (1999) Life, Temperature, and the Earth: The Self-Organizing Biosphere. Columbia University Press, New York. reprinted 2002 as paperback with update.

Schwartzman DW, Caldeira K (2002) Cyanobacterial emergence at 2.8 Gya and greenhouse feedbacks. Geological Society of America Abstracts with Programs Annual Meeting, 34(6): 239–240.

Schwartzman D, Lineweaver CH (2004) Precambrian Surface Temperatures and Molecular Phylogeny. In: Norris R, Stootman F (eds) Bioastronomy 2002: Life Among the Stars ASP Conf Series. Vol 213, pp. 355–358.

Shepard C, Drummond A, Ewing G, Miyauchi S, Rodrigo A, Saul D (2003) Rebuilding the SSU rRNA tree of life using sequence/structure information and Bayesian phylogenetic inference. In: Littlechild JA, Paddy HR, Garcia-Rodriguez E, eds, Thermophiles 2003 Book of Abstracts, University of Exeter, p. 13.

Shock EL, McCollom T, Schulte, MD (1998) The emergence of metabolism from within hydrothermal systems. In: Wiegel J, Adams M (eds) Thermophiles: the keys to Molecular Evolution and the Origin of Life? Taylor and Francis, Chap 5, pp. 59–76.

Stetter, K.O. remarks at Thermophiles 2003 conference held at the University of Exeter 2003.
Szathmary E, Maynard Smith J (1995) The major evolutionary transitions. Nature 374: 227–232.
Tehi M, Franzetti B, Maurel M-C, Vergne J, Hountondji C, Zaccai G (2002) The search for traces of life: the protective effect of salt on biological macromolecules. Extremophiles 6: 427–430.
Tong K-L, Hong X, Wong, J T-F (2003) Rooting the tree of life with tRNA genes. In: Littlechild JA, Paddy HR, Garcia-Rodriguez E, eds, Thermophiles 2003 Book of Abstracts, University of Exeter, p. 122.
Valley JW, Peck WH, King EM, Wilde SA (2002) A cool early Earth. Geology 30: 351–354.
Vogel G (1998) Tracking the history of the genetic code. Science 281: 329–331.
Waters E, Hohn MJ, Ahel I, Graham DE, Adams MD, Barnstead M, Beeson KY, Bibbs L, Bolanos R, Keller M, Kretz K, Lin X, Mathur E, Ni J, Podar M, Richardson T, Sutton GG, Simon M, Soll D, Stetter KO, Short JM, Noordewier M (2003) The genome of *Nanoarchaeum equitans*: insights into early archaeal evolution and derived parasitism. Proc Natl Acad Sci USA 100: 12984–12988.
Wachtershauser G (1998) The case for a hyperthermophilic, chemolithoautotrophic origin of life in an iron-sulfur world. In: Wiegel J, Adams M (eds) Thermophiles: the keys to Molecular Evolution and the Origin of Life? Taylor and Francis, London, pp 47–57.
Wang, DY-C, Kumar S, Hedges, SB (1999) Divergence time estimates for the early history of animal phyla and the origin of plants, animals and fungi. Proc R Soc London B 266: 163–172.
Woese CR (1987) Bacterial evolution. Microbiol Rev 51: 221–271.
Yamagishi A, Watanabe K, Iwabata H, Ohkuri T, Yokobori S (2003) Experimental test: if the common ancestor was a hyperthermophile? In: Littlechild JA, Paddy HR, Garcia-Rodriguez E, eds, Thermophiles 2003 Book of Abstracts, University of Exeter, p. 125.

17 Entropy and Gaia: Is There a Link Between MEP and Self-Regulation in the Climate System?

Thomas Toniazzo[1], Timothy M. Lenton[2], Peter M. Cox[1], and Jonathan Gregory[1]

[1] Hadley Centre for Climate Prediction and Research, Fitzroy Road, Exeter, Devon, EX1 3PB, UK
[2] Centre for Ecology and Hydrology, Bush Estate, Penicuik, Midlothian, EH26 0QB, UK

Summary. The Gaia hypothesis posits that the Earth's climate is self-regulating, while the maximum entropy production (MEP) principle suggests that the climate system self-organizes in a state of maximum entropy production due to turbulent dissipative processes. We explore the relationship between the two by applying MEP to a toy model based on Daisyworld in which the temperature-albedo feedback is dependent on the heat transport rates within the system. We initially assume that the dynamical response of the climate system to differential radiative heating is to create heat fluxes such that a steady state satisfying a maximum entropy-production (MEP) condition is obtained. The resulting system, which does not depend on free parameters, turns out to be thermostatic and to favour the existence of two, but not several, daisy species simultaneously. Furthermore, it maximizes the range of luminosity over which daisies exist, that is, the lifespan of Daisyworld. However, if the daisy coverage is assumed to adjust more slowly than the heat fluxes, the range of habitation is narrowed. Imposing a sinusoidal forcing allows more than two species to coexist, but only occasionally and not to a significant extent.

17.1 Introduction

The Gaia hypothesis (Lovelock 1972, Lovelock and Margulis 1974) emerged from the realisation that Earth's atmospheric composition is in an extreme state of thermodynamic disequilibrium (low entropy), and that this is in turn a product of the presence of myriad non-equilibrium life forms maintaining a highly ordered (low entropy) state (Lovelock 1975). In its later incarnation, the Gaia theory proposed that atmospheric composition and climate are self-regulated by the whole surface Earth system of life plus its abiotic environment ("Gaia"). Self-regulation can be defined in terms of resistance to external forcing or resilience to perturbation (Lenton 2002). Here we study self-regulation of the climate in response to gradually changing external forcing (increasing solar luminosity) in the presence of biotic feedbacks when the MEP paradigm is adopted.

Paltridge (1975, 1978, 1981) proposed that, in analogy to simpler dynamical systems that are subject to a given forcing, the response of the climate system to heterogeneous solar input is to maximize its internal entropy production (and thus maximize the entropy export) due to turbulent dissipation associated with heat transport. This idea is referred to as the maximum entropy production (MEP) principle, and is reviewed elsewhere (O'Brien and Stephens 1995, Ozawa et al. 2003, also Kleidon and Lorenz, this volume). Its applicability to planetary atmospheres and the the Earth is reviewed and discussed in Kleidon and Lorenz (this volume) and in Lorenz (this volume), while Dewar (this volume) provides a theoretical framework.

To investigate the possible effects of large-scale biotic feedbacks, we focus on a simple albedo mechanism, which operates indirectly on the entropy production of the climate through the amount of solar energy absorbed locally. We need thus to distinguish between total planetary entropy export and internal generation of entropy within the system. The Earth's total planetary entropy export is due entirely to radiation, and includes the thermalization of high-energy solar radiation into equilibrium black-body infrared radiation. However, as long as non-linear radiative processes, such as photochemical processes, can be neglected, internal entropy production is nearly independent of the nature of the heating source, but only on its distribution, and is due mainly to turbulent dissipation by the atmospheric and oceanic circulation. Theory and observations suggest that the statistically preferred state can be found by maximizing internal, material entropy production only.

Absorption of radiation is essentially a linear process (Ozawa et al. 2003). Feedback mechanisms can tie albedo and absorption together, via an albedo dependence on temperature. Such feedback may be abiotic or biotic. Well-known abiotic feedbacks include snow/ice-albedo positive feedback, and cloud cover-temperature negative or positive feedbacks. Biotic feedbacks include taiga/tundra-albedo positive feedback, vegetation-water cycle positive feedback, vegetation-cloud cover negative feedback, and the feedback between marine dimethyl sulphide (DMS) production and cloud albedo, the sign of which is uncertain (see also Kleidon and Fraedrich, this volume).

Here we consider whether there is a link between maximization of entropy production due to heat transports (i.e., mechanical and thermal dissipation) within the climate system and self-regulation of the climate system. As a first attempt to address this issue we refer to the simplest 'Gaian' paradigm available: Daisyworld.

17.2 Daisyworld

The Daisyworld model (Watson and Lovelock 1983), describes a multi-component system, forced by a prescribed, homogeneous solar radiation field, with the capacity to vary its albedo and thus the amount of heat taken up. Within a range of solar forcing, the system has built-in feedback mechanisms that keep the surface temperature close to a pre-defined value. This is somewhat analogous to chemical buffer solutions. The concentrations, or amounts, of the different components (e.g., of the carbonate ion in a sodium bicarbonate solution, or the black daisies in Daisyworld) depend on the quantity on which the feedback operates (the pH in the buffer, the temperature in Daisyworld). Pushing the analogy further, in Daisyworld, the "reaction rate" is prescribed, taking the form of a heat exchange rate between the different components of the system. It is this heat exchange rate that makes the feedback operative, and defines the range of external forcing within which it does operate. The analogy, however, ends here. The different daisy species of Daisyworld interact indirectly through the mediation of heat transport, which must be provided by a dynamical response of the climate system. Furthermore, while the near-equilibrium properties of chemical solutions with generic parameters are well known, those of the climate system are not. If the system "chose" to transport no heat, or to guarantee a uniform local temperature everywhere, Daisyworld would not work. Indeed, if poleward heat transport were not taking place in the real Earth climate, daisies and other species would have a pretty hard time almost everywhere.

In order to determine the forcing, and thus the heat exchange rates and local temperatures obtained, most existing MEP studies (Paltridge 1975, Paltridge 1978, Lorenz et al. 2001) prescribe not the external solar radiation field, but the actual heat locally taken up by the climate system, i.e., the local albedo. Here we apply the MEP principle to the Daisyworld system, such that none of the albedo values or heat transport rates have to be prescribed. The surface temperature follows just from the intrinsic properties of the feedback and from the magnitude of the solar luminosity. Moreover, the natural formulation of the equations is such that it is not necessary to assume a given number of species. In principle, any number of species is allowed, and the actual number corresponding to the steady-state solution for a given external radiative field is determined as the one compatible with the MEP principle.

Much of the work described herein was undertaken independently of, and at a similar time to, an existing study (Pujol 2002). Our formulation of the problem in Sect. (17.3) is, however, more general, while the results for a two-daisy system (17.4) are equivalent to those of Pujol (2002) given certain caveats. In the subsequent sections we consider what happens with multiple components/daisies (17.5), with saturating growth responses (17.6), with two boxes that represent equatorial- and polar-regions and exchange heat (17.7), and when the daisies are assumed to adjust more slowly than the climate system (17.8). We close with a discussion in Sect. 17.9.

17.3 Model Formulation

The system is determined by three sets of equations. The first set follows Watson and Lovelock (1983), and describes the biological equilibrium between growth rate and death rate as determined by the competition for resources and by the physiological properties of each daisy species. A key point is that the growth rate of each daisy species is assumed to be proportional to the area of free soil, α_g. The growth equation for species i is:

$$\frac{d \ln \alpha_i}{dt} = \alpha_g \beta_i (T_i) - \gamma_i \,, \quad (i = 1, 2, \ldots, n) \tag{17.1}$$

where $\alpha_g = 1 - \sum_i \alpha_i$, α_i is the fractional area occupied by daisy species i, $\beta_i(T)$ is a function of temperature characterizing the growth rate of daisies of species i, γ_i is the death rate of that species, n is the number of available daisy species.

The functions β_i are assumed to peak at a certain temperature $T_{0,i}$ and reduce to zero (or to negligible values) outside an interval of width $2T_{s,i}$ centred on the peak. We assume:

$$\beta_i (T) = \beta_{0,i} \left[1 - \left(\frac{T - T_{0,i}}{T_{s,i}} \right)^2 \right] , \quad (i = 1, 2, \ldots, n) \,. \tag{17.2}$$

The detailed form of β_i is relatively unimportant, especially when (in thermodynamic units) $T_{s,i} \ll T_{0,i}$, since the daisies are present in significant amounts only near their optimum temperature $T_{0,i}$, where an approximate form (17.2) always holds. In a steady-state, (17.1) can be satisfied only if:

$$|T_i - T_{0,i}| \leq T_{s,i} \sqrt{1 - \gamma_i / \beta_{0,i}} \,. \tag{17.3}$$

In the following, we restrict our attention to steady states, so that all time derivatives will be assumed to vanish. In order to make use of the MEP constraint, we must assume that the equilibration of the daisy system is much faster than any characteristic dynamical time scale of the climate system. Additionally, we assume (quite reasonably) that the radiative forcing, T_r, changes slowly compared to the evolution of the daisies.

Equation (17.1) can be written in the more convenient form:

$$b_i (\theta_i) = \phi \qquad (i = 1, 2, \ldots, n) \tag{17.4}$$

where $b_i = \beta_i / \gamma_i$ is the normalized growth function for species i, and $\theta_i = T_i / T_{0,i}$. The quantity $\phi = 1/(1 - \sum_i \alpha_i)$ is the inverse of the bare-soil fraction, and it determines the local temperatures of the existing daisy species, since

$$T_i = T_{0,i} b_i^{-1} (\phi) \qquad (i = 1, 2, \ldots, n) \,. \tag{17.5}$$

The second set of equations for the MEP-Daisyworld model expresses the steady-state energy balance of the system. Assuming grey-body spectra:

$$\alpha_i \left(T_i^4 - N_i T_r^4\right) = Q_i \qquad (i = 1, 2, \ldots, n)\,, \tag{17.6}$$

$$T_g^4 - N_g T_r^4 = -\phi \sum_i Q_i \tag{17.7}$$

Here, the subscript g refers to the bare soil; $N_i = 1 - A_i$, where A_i is the albedo of the daisies of species i, and similarly $N_g = 1 - A_g$; T_r is the effective (or equivalent black-body) temperature of the solar radiation field.

The quantities Q_i are heat fluxes per unit area (divided by the Stefan-Boltzmann constant), which it is assumed the climate system provides to each daisy species. In the original model (Watson and Lovelock 1983) they took the explicit form $Q_i = \alpha_i q(A - A_i)$, where A is the average albedo as resulting from the fractional area covers α_i of the daisy species, and q is a free parameter, which must lie in the range $0 \leq q \leq T_r^4$ in order to satisfy Kelvin's 2nd law of thermodynamics. The range of T_r over which the albedo feedback operates (if at all), and the number of daisy species allowed by the system at a certain T_r, all depend on the adopted value of q.

In the present formulation, the heat fluxes Q_i are treated as independent quantities. Then, by substituting α_i from (17.6) and T_i from (17.7), the definition of ϕ can be written

$$\frac{1}{\phi} = 1 - \sum_i \frac{Q_i}{\left[T_{0,i} b_i^{-1}(\phi)\right]^4 - N_i T_r^4} \tag{17.8}$$

The solution of this equation, $\phi = \phi(\underline{Q})$, falling within the interval $1 \leq \phi \leq \max_i \{\beta_{0,i}/\gamma_i\}$ characterizes the response of the Daisyworld system to a given set of values of the heat fluxes Q_i. The local temperatures T_i and the fractional area coverages α_i are given by (17.5–17.6). The heat fluxes Q_i are determined as functions of the forcing radiation brightness temperature T_r by a final set of equations which follow from the MEP requirement, i.e., that the entropy production within the system attains a maximum. The entropy production is given by:

$$\dot{S} = \sum_i Q_i \left(\frac{1}{T_i} - \frac{1}{T_g}\right), \tag{17.9}$$

where T_g is given by (17.7), and the MEP conditions can be written:

$$\partial_{Q_i} \dot{S} = 0 \qquad (i = 1, 2, \ldots, n)\,. \tag{17.10}$$

The form of (17.6) gives a parameterization of the heat transports, with respect to which $\dot{S}$ as given in (17.9) is to be maximized, that is consistent with the application of the MEP principle. The entropy production is a function of quantities defining the heat transfer rates between different components of the system under a fixed external forcing, i.e., given an external energy input into the system. The novelty of the present system is that it can control the external energy flux by changing the fractional area coverage of the daisies and thus the average albedo. The heat-flux rates, though, do not depend on the fractional areas.

The possible values of Q_i are constrained by two conditions. First, the range of values for the temperature T_i, obtained from (17.6) when α_i varies between 0 and $1 - \min_i(\gamma_i/\beta_{0,i})$, should overlap with the interval defined by (17.3). Second, from Kelvin's second law of thermodynamics, the temperature of species i cannot be higher than the equilibrium temperature for the species with the largest available albedo when no heat is exchanged, and cannot be lower than the corresponding case for the lowest available albedo.

When one of the Q_i is equal to zero (say, for $i = i_0$), (17.6) has two solutions. One is $\alpha_{i_0} = 0$, and it is also found by solving (17.8); the other is $T_{i_0} = N_{i_0}^{1/4} T_r$. For this second case, ϕ is obtained from (17.5) for $i = i_0$, the remaining temperatures from the same equation with $i \neq i_0$, the values of α_i for $i \neq i_0$ from (17.6), and finally $\alpha_{i_0} = \phi^{-1} - \sum_{i \neq i_0} \alpha_i$. Similar considerations apply if more than one of the Q_i's is zero. A solution with $Q_i = 0$ and $\alpha_i \neq 0$ does not represent a MEP solution. For such solutions, a set of infinitesimal variations δQ_i, $i = 1, 2 \ldots$, can always be found such that the entropy production increases. Therefore, such solutions are not considered further in the MEP context.

In view of the above, it is convenient to simplify the computation by using (17.6) to change variables to the fractional areas α_i. Given a vector of values $\underline{\alpha}$ for the fractional areas, one can calculate ϕ, and thus the temperatures $\underline{T}$ from (17.5), the fluxes $\underline{Q}$ from (17.6), and finally formally maximize $\dot{S}$ with respect to $\underline{\alpha}$. Since $Q_i \approx 0$ as $T_i \to N_i^{1/4} T_r$, the transformation is smooth. Thus, the $\dot{S}$ surface is mapped faithfully from one variable space into the other, single maxima corresponding to single maxima. This is not guaranteed if a different parameterization is chosen for the heat fluxes, for example, following Watson and Lovelock (1983), $Q_i = \alpha_i q(A - A_i)$, and allowing q to be a function of daisy species (i.e., $q = q_i$), leads to multiple local entropy maxima. In the present formulation, instead, the maximum seems to be unique in continuum space (see below). This property may be relevant when the global maximum changes location discontinuously, since the implied "phase transition" of the climate system will be of a different order.

A complication arises from the fact that the solution for a given $\underline{Q}$ or a given $\underline{\alpha}$ is not unique when the growth-rate functions β have multiple-valued inverse, as those defined in (17.2). In particular, whenever the growth function has a maximum, (17.5) has, for each i, two solutions, one lying above and the other below the peak-growth temperature $T_{0,i}$ (with the exception of the case $T_i = T_{0,i}$). As a result, in an N-daisy "world" there are, for each set of areas $\underline{\alpha}$, 2^N different solutions, i.e., 2^N sets of temperatures $\underline{T}$ and of corresponding heat fluxes $\underline{Q}$. We may express this multiple-valued-ness by assuming that the solution depends on an additional, discrete variable, say k, with $k \in 1, 2, \ldots, 2^N$. Suppose the system is found in one particular state $\{\underline{\alpha}, \underline{T}, \underline{Q}\}$. Statistical fluctuation in the system will ensure that the local neighbourhood in the continuum heat flux (or area-) space is "explored", and thus an extremal principle formulated in these variables is justifiable. The up to 2^N solutions to (17.5), however, are not continuously connected to one another. On the one hand, if the level of fluctuations within the system is such as to bring it from one value of k to the other, transients will be significant, and the equilibrium states will not be sufficient to characterize the solution. If, on the other hand, fluctuations are small compared to the phase-space gap between solutions at different k, there is no guarantee that the global maximum of entropy production is always achieved. Therefore, conformity to the MEP principle can be interpreted in the sense that $\dot{S}$ is a maximum in heat-flux space, while it may not be a maximum with respect to the discrete variable k.

17.4 Two-Component System

The simplest system displaying significant homeostatic properties is that allowing a maximum of two daisy types simultaneously. The intervals within which each daisy type can be expected to cover a non-zero area are once again found from (17.3) and (17.6). If the subscript 'w' ('b') indicates the daisy species with the higher (lower) albedo, the brightness temperature of the radiation field must lie in the range

$$\frac{\max\left(T_{0,i} - T_{s,i}\sqrt{1 - \gamma_i/\beta_{0,i}}\right)}{N_b^{1/4}} \leq T_r \leq \frac{\min\left(T_{0,i} + T_{s,i}\sqrt{1 - \gamma_i/\beta_{0,i}}\right)}{N_w^{1/4}} \quad (17.11)$$

when both daisy species are present. The max and min functions arise because the additional condition $T_b \geq T_w$ holds. For at least one daisy species to be present, T_r must be larger than the smaller of the expressions $\left(T_{0,i} - T_{s,i}\sqrt{1 - \gamma_i/\beta_{0,i}}\right) \Big/ N_i^{1/4}$, and similarly for the upper bound.

An example of a MEP Daisyworld solution is shown in Fig. 17.1. The entropy production, fractional areas α_i, daisy temperatures T_i, and heat fluxes Q_i are plotted for each combination of root choice in (17.5). The MEP solutions provide heat fluxes such that the range of T_r within which daisies can survive is maximal. In a q-prescription as in Watson and Lovelock (1983), this range is significantly reduced with respect to the MEP case. An example

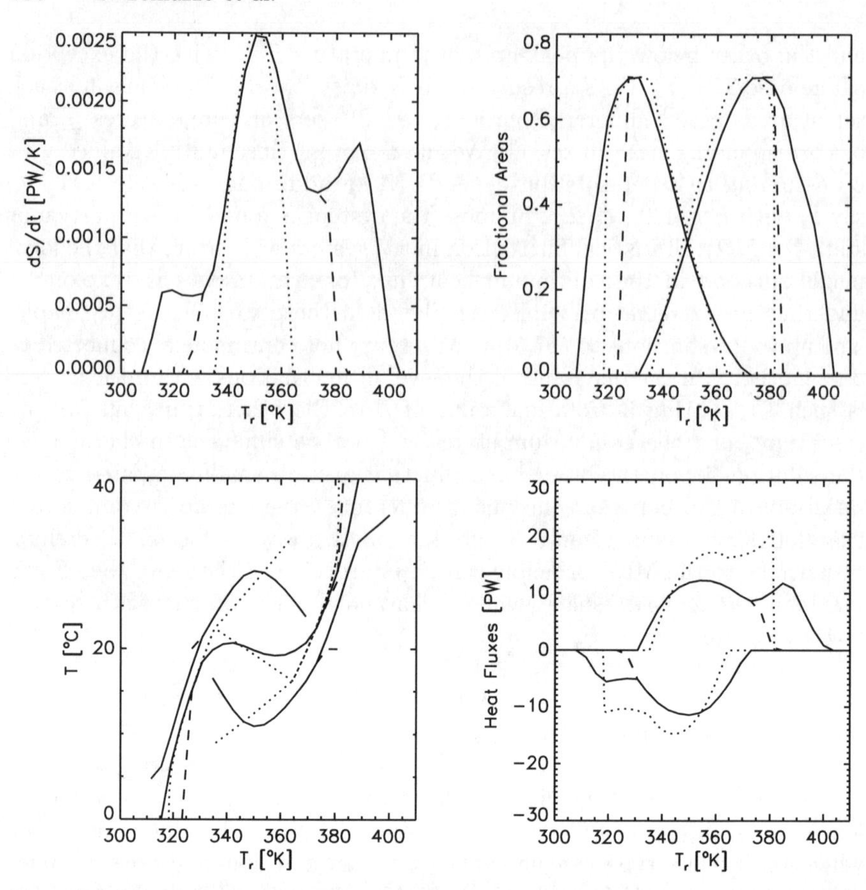

Fig. 17.1. Two-species "Fast daisy" MEP system (*solid and broken lines*) compared to the original Daisyworld solution (*dotted lines*). The *broken line* shows the MEP solution with $k = 0$; the *solid line* is the overall MEP solution among the continuous (constant-k) solutions. Panels show entropy production (*upper left*), fractional area coverage for the two species (*upper right*), the temperatures for each species and average temperature (*lower left*), and the heat fluxes towards each species (*lower right*). The parameters of the model are $N_1 = 0.35$, $N_2 = 0.65$, $T_{0,1} = T_{0,2} = 20°\mathrm{C}$, and $T_{\mathrm{s},1} = T_{\mathrm{s},2} = 20°\mathrm{C}$

of such a solution is shown together with the MEP solutions in Fig. 17.1. The MEP closure implies that both the range over which daisies exist and that over which both species exist are increased. Note, however, that the MEP solutions imply less comfortable temperatures for the daisies, and hence at some times a smaller total daisy area coverage. Within the MEP constraint, the climate system does not support the heat fluxes required for the daisies to spread further. Thus, in the present model framework, homeostasis arises not as a consequence of self-regulation of the daisies alone, but as a preferred state of the overall climate system when it comprises the daisies as a sub-system.

17.5 Multi-component System

The present model is clearly formulated for an arbitrary, but finite, number of species. Among those, it selects the species that are supported by the dynamical system "responsible" for providing the heat fluxes. In some sense, therefore, the model is parameter-free. The functional form of the growth-functions β_i does not need to be the same for each species either. Obviously, it is practically impossible to provide a "complete" input to the numerical maximization routine (even within the restrictive model assumptions). In fact, the basic character of the solutions can be investigated using only a small number of species. In a three species system we find that there are no continuous solutions in which the three species coexist. The overall MEP solution in k-space also only supports two species: those with maximum albedo contrast. This even reaches the paradoxical situation when the species whose optimum temperature $T_{0,i}$ is at its effective radiation temperature $N_i^{1/4}T_r$ is not supported. This species would tend to occupy a large area without exchanging heat with the surroundings, a situation evidently far from MEP. Even if the species with intermediate albedo is a "super species" with a very wide range of operating temperatures, when the other two species can manage to survive, it tends to be replaced. Thus the three species system tends to behave as a "piecewise two-species" system. Once again, whether any such condition is achieved depends on whether there is an amount of fluctuations sufficient to bring the system from one MEP branch to the other, or not.

17.6 Saturated Growth

The ambiguity in the solution is avoided if one chooses monotonic growth functions β_i, which saturate to γ_i either for small or for large values of the temperature. Such functions may describe non-biological components like sea-ice, or rising sea level. For the purpose of illustrating the case, we assume a form

$$\beta_i = \max\left(\beta_{0,1} \tanh\left(2[(T - T_{0,i})/T_{s,i} + 1]\right), 0\right) \tag{17.12}$$

where $T_{s,i}$ may be positive (saturation at high T) or negative (saturation at low T), and $T_{0,i} - T_{s,i}$ determines where the function goes to zero, i.e., the constants are chosen such that $T_{s,i}$ has a similar width-determining role as in (17.2).

Figure 17.2 illustrates a 2-species case for the same parameters as in Fig. 17.1, except that T_s has the opposite sign and a smaller magnitude for the light species. The similarity is obvious, and stems simply from the fact that the $\beta_i(T)$ curves are very similar in both cases when one restricts the attention to the sloping parts at $T > T_0$ for the dark and $T < T_0$ for the light species. The physical meaning of this is that regulation originates from the

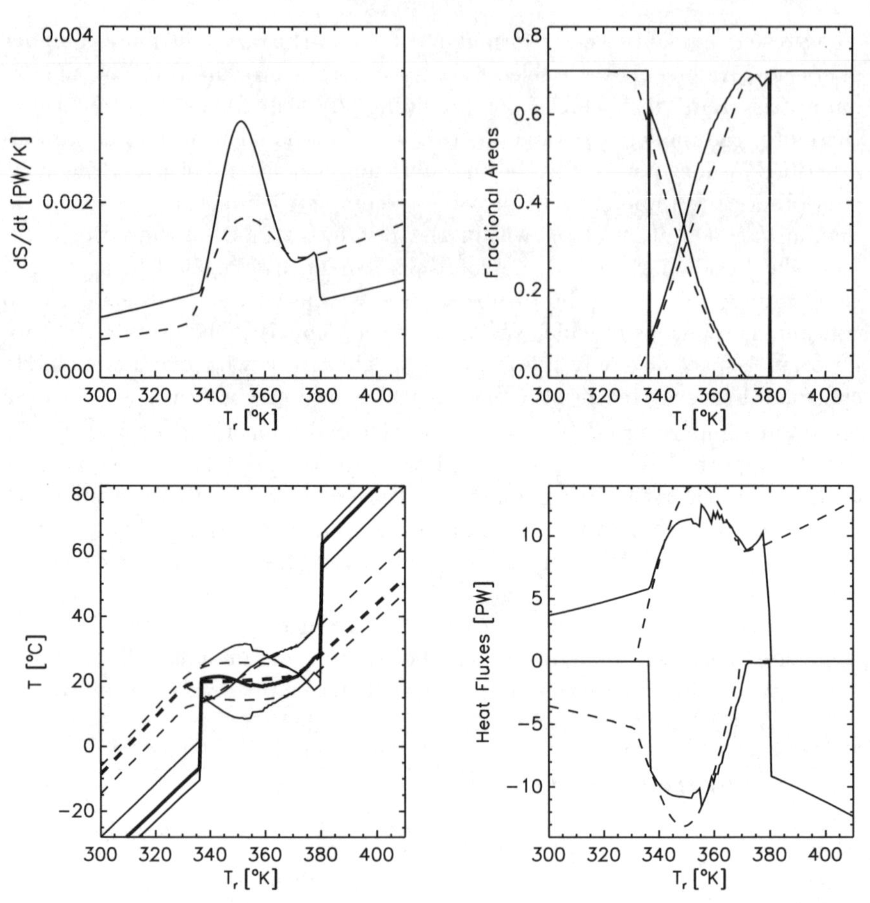

Fig. 17.2. Two-species saturated-growth "Fast daisy" MEP system. The panels are as in Fig. 17.1, but the *solid lines* now show the case in which the higher-albedo species ($N_1 = 0.35$) is "cold-loving" (i.e., its growth function saturates to 1 at low temperatures) and the lower-albedo species ($N_2 = 0.65$) is "warm-loving" (growth-rate 1 at high temperatures), while the *broken lines* show the opposite case ("cold-loving" dark species and "warm-loving" bright species). The other model parameters are $T_{0,1} = T_{0,2} = 20°\text{C}$ in both cases, $T_{\text{s},1} = -10°\text{C}$, $T_{\text{s},2} = 20°\text{C}$ in the first case, and $T_{\text{s},1} = 10°\text{C}$, $T_{\text{s},2} = -20°\text{C}$ in the second case. Note that in the first case the area coverage of both species has jumps marking the transitions between the two-species "world" and a single-species "world". The *thicker lines* in the temperature panel indicate the average (planetary) temperature, while the other three lines show the local temperatures for each species and for the bare soil

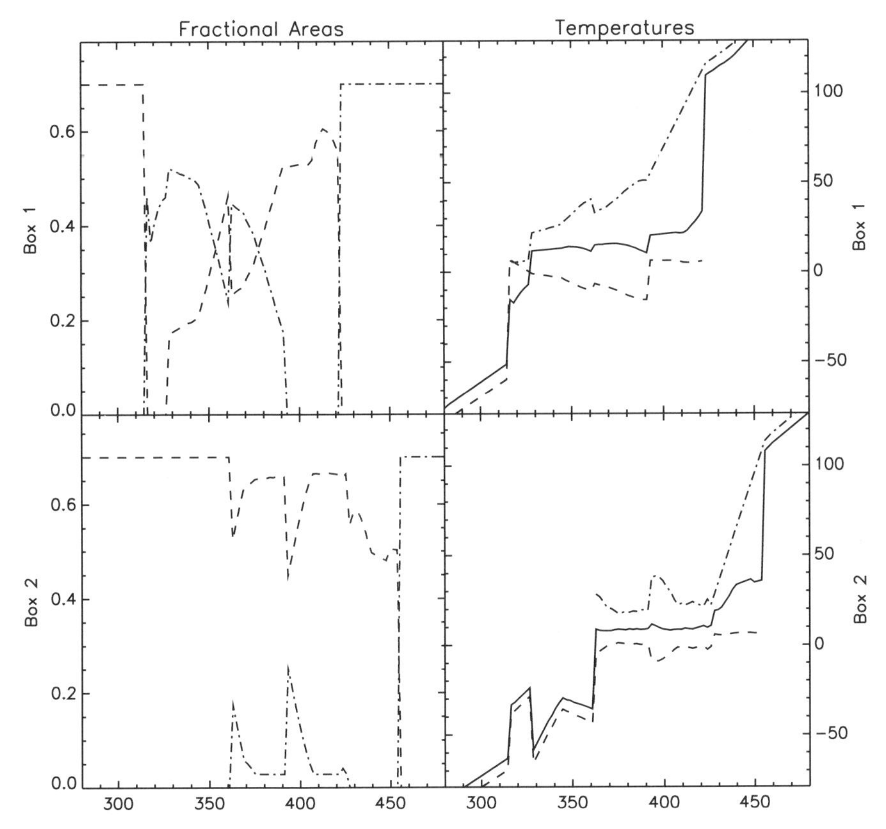

Fig. 17.3. Two-species, two-box "Fast daisy" MEP system with saturating growth-functions. Box 1 corresponds to the "tropical" box 'E', and Box 2 to the "subpolar" box 'P'. *Dashes lines* refer to the bright species ($N_1 = 0.2$), *dot-dashed lines* to the dark species ($N_2 = 0.8$). *Solid lines* in the temperature panels indicate the ("zonal") average temperatures. The other model parameters are $T_{0,1} = 0°\mathrm{C}$, $T_{\mathrm{s},1} = -10°\mathrm{C}$, $T_{0,2} = 20°\mathrm{C}$, $T_{\mathrm{s},2} = 20°\mathrm{C}$. The case presented is similar to that shown as *solid lines* in Fig. 17.2 ("cold-loving" bright species and "warm-loving" dark species)

coupling of two positive feedback mechanisms, and that MEP favours this coupling. (A similar example with three species is not worth showing, since the intermediate species seems to always be cut out when (17.12) is used.)

17.7 A Two-Box Model

An interesting modification of the homogeneously forced system is to allow two subsystems to be forced by radiation fields with different brightness temperature, couple them once again via heat exchange, and maximize the entropy production arising from this coupling. Equation (17.6) is replaced by

$$\alpha_i \left(T_i^4 - N_i T_r^4 - Q_B\right) = Q_i\,, \tag{17.13}$$

where Q_B is the heat exchanged between the two boxes. $\dot{S}$ is maximized for each box separately for a given Q_B, and then the additional quantity

$$\dot{S}_B = Q_B \left(\frac{1}{T_P} - \frac{1}{T_E}\right) \tag{17.14}$$

is maximized, where T_P ("poles") and T_E ("equator") are the average temperatures of the two boxes, e.g.,

$$T_E = (T_g)_E + \left(\sum_i \alpha_i (T_i - T_g)\right)_E \tag{17.15}$$

We adopted a radiation field of varying intensity and multiplied it by the factor $\pi/6 + \sqrt{3}/4$ for box 'E', and by $\pi/3 - \sqrt{3}/4$ for box 'P', in order to simulate the irradiation impinging on the tropical sector between 30 °N and 30 °S, and that on the remaining sub-polar regions. The same two-species system was assumed to be (potentially) present in either region. Figure 17.3 shows the resulting temperatures and area coverage for the case of saturating β_i's, (17.12). Here species 1 is even more "cold-loving" than before with $T_{0,1} = 0°$ C, $T_{s,1} = -10°$C. Coupled, sharp state transitions occur repeatedly in both boxes.

17.8 Slow Daisies

In the steady-state approach described so far one crucial assumption has to be made: that the daisy population equilibrates on a time-scale much shorter than that needed by the heat fluxes to adjust to changing daisy area coverage. Only under this assumption are the algebraic relationships in (17.4) valid with the equilibrium temperatures derived from entropy-production maximization.

This assumption is of course arbitrary. The atmospheric adjustments responsible for most of the heat transport in temperate land areas occur on time-scales of the order of a month to a year; deep ocean processes can take much longer, but with the exception of the North Atlantic the bulk of the ocean heat transport is related to gyre circulations and shallow wind-driven overturning cells, with time-scales of the order of a decade. It is not straightforward to think of an example of planetary-scale biota evolution that happens on timescales much shorter than this, especially when populations (i.e., area coverages in DW) are small.

Moreover, a second, possibly stronger assumption is also implied in the above treatment of the DW problem. MEP for Daisyworld involves two maximizing assumptions, not only that (i) the heat transport adjusts to MEP for given external heating rates (i.e., albedos) but also that (ii) the heating rates, or daisy fractional covers, adjust to MEP as well. While the first assumption represents the original application of MEP, it is not clear to us how justifiable the second assumption (and its combination with the first) is. For daisy areas to be regarded as a variable with respect to which EP is maximized (under the daisy-growth steady-state assumption), the climate system must have the ability not only to pick the MEP trajectory among those allowed when the external heating rates (albedos) are fixed (i.e., the original MEP principle for steady-states), but also among those different MEP states with different heating rates for which the actual entropy production is greatest.

To address these issues, a second set of calculations have been performed in which the above assumptions are relaxed in favour of another approximation, namely that the MEP-adjustment time-scale be much shorter than the daisy growth time-scale. Entropy production is given by (17.9), the temperature of each daisy species T_i from (17.6), and the temperature of daisy-free land T_g from (17.7). Entropy production is now maximized with respect to all T_i's (or, equivalently with respects to all Q_i's), with areas α_i kept fixed. This leads to the set of equations:

$$\frac{\partial \dot{S}}{\partial t} = \alpha_i T_i^3 \left[\frac{1}{T_i} \left(3 + \frac{N_i {T_r}^4}{{T_i}^4} - \frac{1}{T_g} \left(3 + \frac{N_g {T_r}^4}{{T_g}^4} \right) \right) \right] = 0 \qquad (17.16)$$

The solution method consists in taking an initial guess for

$$C = \theta_g \left(3 + N_g \theta_g^4 \right) , \qquad (17.17)$$

where $\theta_g \equiv T_r / T_g$, and solving the equation:

$$\theta_i \left(3 + N_i \theta_i^4 \right) = C \qquad (17.18)$$

for each i, to obtain $\theta_i = \theta(N_i, C)$. From the θ_I's, $\theta_g = \theta(C)$ is then computed. Equation (17.17) for C is solved via bisection. Because $0 < N_{i,g} < 1$ and $0 < \theta_{I,g} < 1$ in (17.18) and (17.17), the solution is guaranteed to exist, and to be unique. The Hessian of $\dot{S}$ can be shown to be negative and hence the solution represents a unique maximum for $\dot{S}$ in T (or Q) space.

Thus for each set of values of $\{\alpha_i, i = 1, ..., n\}$ we have derived, under the MEP steady-state assumption, a set of internal heat fluxes (Q_i) and hence equilibrium temperatures (T_i) for the different daisy species. Equation (17.1) can now be integrated forward in time to study the evolution of the daisies.

An example is shown in Fig. 17.4 and compared with a steady-state ("fast daisy", or FD) case, with the same parameters. For the time-dependent ("slow daisy", or SD) case shown, the time-scale over which the solar forcing (or its effective brightness temperature T_r) varies is set at 10^6 times the inverse daisy maximum growth rate. Therefore, except for very small values of α_i (not distinguishable in Fig. 17.4) the SD system is in a steady state to a very good approximation. (A "seed" value for α has to be specified in the calculation for daisies to grow. This was taken at 10^{-16}; whenever an α_i drops

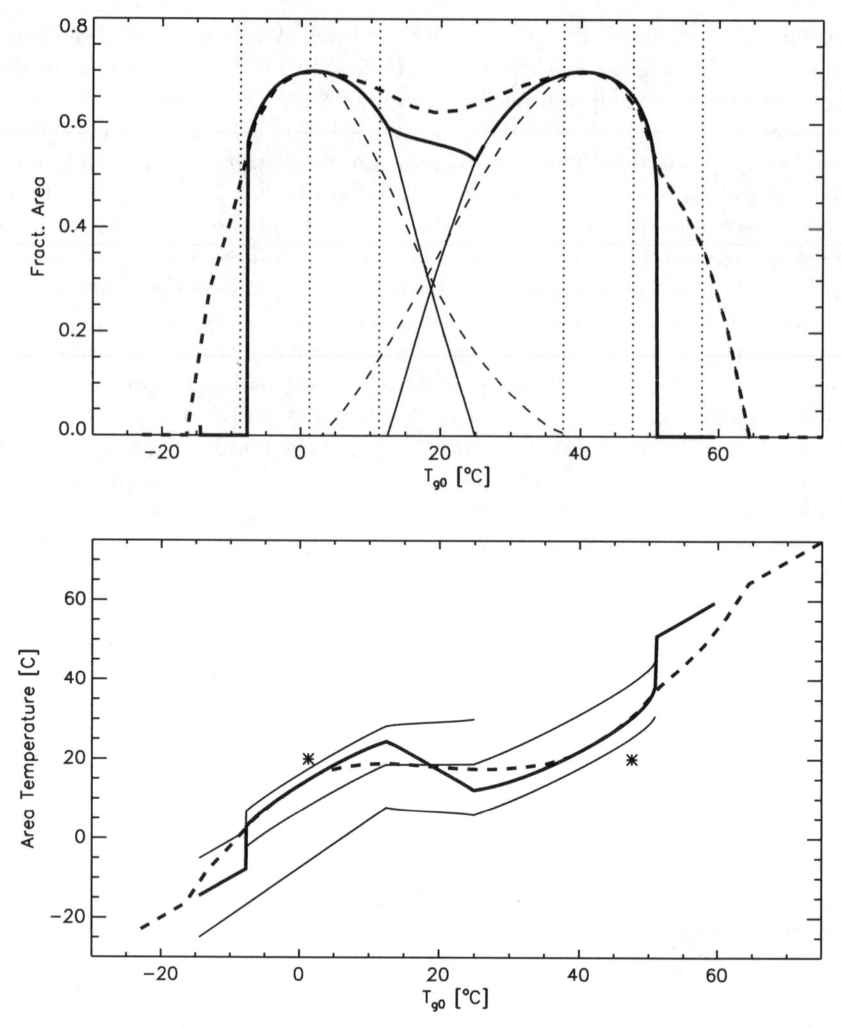

Fig. 17.4. "Fast daisy" (FD) vs. "slow daisy" (SD) MEP models. Results are shown from two calculations with $n = 2$ and $N_1 = 0.35$, $N_2 = 0.65$, optimum temperatures $T_0 = 20°\mathrm{C}$ and a tolerance range $T_\mathrm{s} = 20°\mathrm{C}$, and a death rate $\gamma = 0.3$ of the maximum growth rate. *Solid lines* are for the SD model, *broken lines* for the FD model. The *top panel* shows the fractional area coverage for each species (*thin lines*), and the total daisy area coverage (*thick lines*) in the two models. The vertical *dotted lines* indicate the bare soil temperature ranges corresponding to daisy growth temperatures if internal heat fluxes were zero. The *bottom panel* shows daisy temperatures and bare soil temperature for the SD model (*thin lines*), and planetary average temperatures (*thick lines*) for both models. The two asterisks indicate the optimum temperatures for the two daisy species in the case of no internal heat fluxes

to or below that value, it is not taken account of for EP maximization, and is reset to the seed value. Consistently, only daisy temperatures for α_i greater than seed are shown in Fig. 17.4.)

Some differences between the two cases are noteworthy. The SD case has different regulation properties than the FD case. Entropy production (not shown) is always smaller. Daisy temperatures are generally less optimal for growth and consistently the total area covered by daisies is generally smaller. This is more pronounced under more extreme external forcing conditions. In particular, black daisies do not grow until the bare soil temperature falls within their tolerance window. For white daisies, the environment actually turns hostile at higher solar forcing: the heat conveyed to them from bare-soil areas kills them off while the solar effective brightness temperature is still almost 8 degrees lower than what they could tolerate if there were no exchange heat fluxes. This property of the SD problem is opposite to that of the FD case, where the system "chooses" to keep some daisies alive by reducing heat fluxes because that is still the way to allow for some entropy production. The system with the time-evolving daisies of course does not "know" that it is killing them and just transfers the amount of heat that maximizes EP when the area covers are fixed.

Also the coexistence of daisy species is less likely in the SD case. Some heat flux is provided by the system to the white daisies at low values of T_r (their area fraction coverage is consistently above seed value), but they cannot compete with the black daisies for area occupation. That is possible only after the solar effective brightness temperature has moved out of the zero-heat-flux tolerance window for the black daisies. Thereafter the black daisy population declines and the white daisy population starts filling in the bare soil. The period of coexistence of the two species is shorter than in the FD scenario, and the soil occupation fraction during that time is particularly reduced. Note that the timescale over which these changes happen is 5 orders of magnitude larger than the time daisies would need to occupy the soil under optimal conditions: the occupation fraction is small not because the daisies have not had time to grow, but because they are subject to environmental stress. This again is an important difference from the FD scenario.

The diversity question is particularly interesting. Many-species systems do not show any significant difference from the two-species system shown. The daisy species take over one after the other, with relatively short intervals of time when two coexist, one slowly declining and the other slowly gaining ground. This is an intrinsic limitation of the model that envisages albedo as the only feedback mechanism, while not allowing for rapid fluctuations in the heat fluxes. If a fast sinusoidal forcing is superimposed on the trend (Fig. 17.5), more than two species can coexist occasionally. However, at almost all times one species dominates.

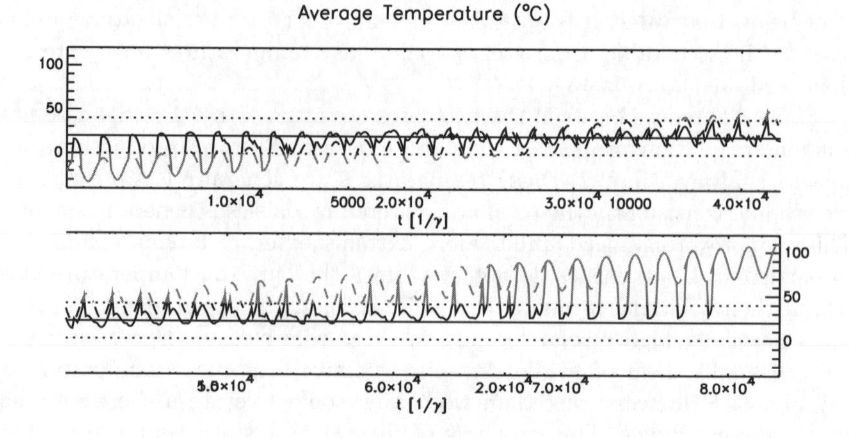

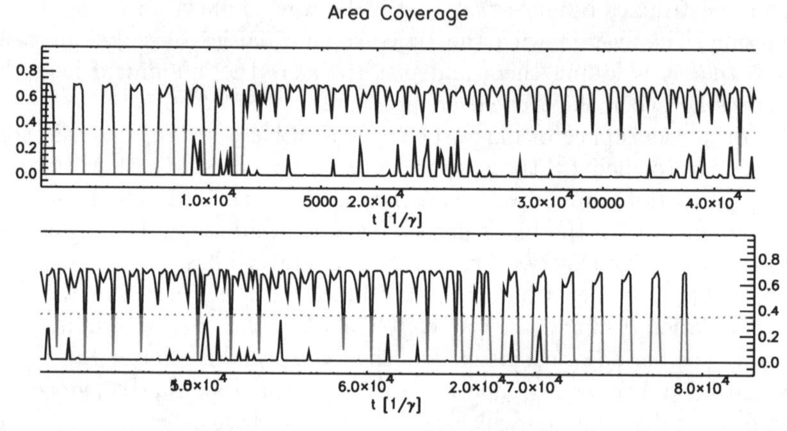

Fig. 17.5. "Slow daisy" MEP Daisyworld model with $n = 6$ and a variable forcing applied. Parameters are as in Fig. 17.1 except that daisy albedos are equal to 0.2, 0.3, 0.4, 0.6, 0.7, and 0.8. The *top panel* shows the average (planetary) temperature with (*solid line*) and without (*broken line*) daisies. Values outside the growth range for daisies are plotted in grey. The abscissa is the elapsed time in units of the inverse daisy death rate. The *bottom panel* shows the total area coverage of all daisy species (*upper line*) and the area occupied by daisy species other than the most abundant one (*bottom line*). Most of the time only one species exists, and almost all the time only one species is dominant in this model

17.9 Discussion

Strictly speaking, the MEP principle is formulated only for dynamical systems with a well-defined accessible phase-space. We have formulated a series of model systems in which there is strong temperature-albedo feedback and entropy production due to heat transports is maximized as a constraint. All of the systems have thermostatic properties and tend to favour the existence of two, but not several, daisy species simultaneously. If the biological populations are assumed to adjust faster than the heat transports, the system maximizes the range of luminosity over which daisies exist (the lifespan of the biosphere). This represents optimal self-regulation in terms of maintaining habitability in response to external forcing. However, if the populations are assumed to adjust more slowly than the heat fluxes, the range of habitation is narrowed. If a sinusoidal forcing is superimposed on the long-term trend, brief periods of coexistence of more than two species occur, but only to a very limited extent.

In model systems based on Daisyworld the total planetary entropy production, which is dominated by radiation absorption, does not show a maximization tendency. In the original model (Watson and Lovelock 1983), the appearance of black daisies lowers planetary albedo, increases solar absorption, and corresponds to a step increase in planetary entropy production. However, as solar luminosity increases and the white daisies gradually take over this amounts to the system counteracting a potential increase in planetary entropy production by maintaining a more constant net absorption flux. When the white daisies disappear this again causes a step increase in planetary entropy production. Thus, there are "phase transitions" corresponding to increases in planetary entropy production, and these are "irreversible", in the sense that they can only be reversed with a significant decrease in solar luminosity or extinction event (i.e., there is hysteresis in the system). Imposing an MEP constraint on heat transport, as we have, removes the regions of hysteresis present in the original Daisyworld, but the system still has a tendency to counteract increasing luminosity by increasing planetary albedo.

Coupling mechanisms between the climate system and the biosphere like that considered in our Daisyworld model can imply that additional degrees of freedom are made potentially available to the climate system on the timescales of adjustment of the ocean-atmosphere system. If so, the MEP principle, as applied here, would need to be modified to account for the additional entropy associated with 'biotic' degrees of freedom, and thermal entropy production would no longer be the (only) relevant quantity to consider. If, on the other hand, the coupling is weak, as e.g., for biota that evolve or grow more slowly than changes in the climate, then we cannot expect to find the daisies in a steady-state. In fact, a general caveat for the application of MEP (in any form) to the climate system is that the conditions and assumptions under which the real climate system may be considered at steady state are not well established.

The Daisyworld studies suggest that while self-regulation of the climate system may be enhanced by maximization of the entropy production due to heat transports, self-regulation bears no clear relationship to total planetary entropy production. This contrasts with a recent simple model of Earth's carbon cycle and temperature. Kleidon (2004) suggests that Earth's planetary albedo is close to a minimum with respect to surface temperature, which exists because of increasing snow and ice cover in response to cooling and increasing (cumulus) cloud cover in response to warming (see also Kleidon and Fraedrich, this volume). A biotic entropy production is defined as the energy flux of respiration divided by the temperature of respiration minus the energy flux of gross primary productivity divided the temperature of radiation used in photosynthesis. This biotic entropy production, like planetary entropy production, is primarily determined by absorbed solar radiation and hence the planetary albedo. If biotic entropy production is maximized (by varying the fraction of biomass that is respired, which is assumed to represent degrees of freedom introduced by the biota), this results in a reduction in atmospheric carbon dioxide as solar luminosity increases and homeostatic regulation of global temperature. In essence, temperature is held close to the value that is assumed to minimize planetary albedo and maximize planetary entropy production (see also Kleidon and Fraedrich, this volume).

Whether a single minimum of planetary albedo exists with respect to temperature is debatable. In the real world, life affects planetary albedo and organisms typically have a peaked growth response to temperature, which could give rise to multiple minima and/or maxima of albedo with respect to temperature. Life can increase planetary entropy production by lowering surface and/or planetary albedo, as do the boreal forest and the black daisies in Daisyworld. However, there are also white daisy analogues in the real world. Phytoplankton that emit dimethyl sulphide (DMS) gas inadvertently increase cloud albedo. Vegetation can enhance low cloud cover through evapotranspiration and the emission of volatile organic carbon compounds that oxidize to form cloud condensation nuclei. Thus it is not clear whether the net effect of life is to increase or decrease planetary albedo. Kleidon (2004) estimates that land vegetation increases planetary entropy production, but a similar analysis is lacking for the marine biota. Kleidon (personal communication) argues, however, that the many ways that the biota affects clouds represent many degrees of freedom, and that is precisely why MEP should apply.

If we accept that given sufficient degrees of freedom, a climate system will tend to adopt a state that maximizes entropy production due to heat transport, the present study and other work (Gerard et al. 1990, Pujol 2002) suggests that this in turn tends to increase the range of solar forcing over which a planet remains habitable. In the Daisyworld studies this result is contingent on the presence of life. However, even in an abiotic model (Gerard et al. 1990), assuming MEP associated with heat transport tends to lower the luminosity threshold for a snowball Earth. If the MEP principle and this tentative link to climate regulation can be generalized it should make us more optimistic about finding life on potentially habitable extra-solar planets.

References

Dewar R (2003) Information theory explanation of the fluctuation theorem, maximum entropy production and self-organized criticality in non-equilibrium stationary states. J Phys A 36(3): 631–641.
Gerard JC, Delcourt D and Francois LM (1990) The Maximum-Entropy Production Principle in Climate Models – Application to the Faint Young Sun Paradox. Q J Roy Meteorol Soc 116(495): 1123–1132.
Kleidon A (2004) Beyond Gaia: Thermodynamics of life and Earth system functioning. Clim Ch: in press.
Lenton TM (2002) Testing Gaia: the effect of life on Earth's habitability and regulation. Clim Ch 52: 409–422.
Lorenz RD, Lunine JI, Withers PG and McKay CP (2001) Titan, Mars and Earth: Entropy Production by Latitudinal Heat Transport. Geophys Res Lett 28(3): 415–418.
Lovelock JE (1972) Gaia as seen through the atmosphere. Atmospheric Environment 6: 579–580.
Lovelock JE (1975) Thermodynamics and the recognition of alien biospheres. Proc Roy Soc London 189: 167–181.
Lovelock JE and Margulis LM (1974) Atmospheric homeostasis by and for the biosphere: the gaia hypothesis. Tellus 26: 2–10.
O'Brien DM and Stephens GL (1995) Entropy and climate. II: Simple models. Q J Roy Meteorol Soc 121: 1773–1796.
Ozawa H, Ohmura A, Lorenz RD and Pujol T (2003) The second law of thermodynamics and the global climate system – A review of the maximum entropy production principle. Rev Geophys: 41, 1018.
Paltridge GW (1975) Global dynamics and climate – a system of minimum entropy exchange. Q J Roy Meteorol Soc 101: 475–484.
Paltridge GW (1978) The steady-state format of global climate. Q J Roy Meteorol Soc 104: 927–945.
Paltridge GW (1981) Thermodynamic dissipation and the global climate system. Q J Roy Meteorol Soc 107: 531–547.
Peixoto JP, Oort AH, de Almeida M, Tomé A (1991) Entropy budget of the atmosphere. J Geophys Res 96(D6): 10981–10988.
Pujol T (2002) The consequence of maximum thermodynamic efficiency in Daisyworld. J Theor Biol 217(1): 53–60.
Rodgers CD (1976) Minimum entropy exchange principle – reply. Q J Roy Meteor Soc 102: 455–457.
Watson AJ and Lovelock JE (1983) Biological homeostasis of the global environment: the parable of Daisyworld. Tellus 35B: 284–289.

18 Insights from Thermodynamics for the Analysis of Economic Processes

Matthias Ruth

Environmental Policy Program, School of Public Policy, University of Maryland, College Park, MD 20742, USA

Summary. The laws of thermodynamics constrain transformation of materials and energy, and thus have implications for economic processes. This paper provides an overview over the uses and applications of concepts from thermodynamics in economic analysis at the level of individual processes and explores potential constraints at larger system levels – the economy as a whole and the ecosystems within which economies are embedded. Specific emphasis will be placed on the ways in which insights from equilibrium and non-equilibrium thermodynamics can be used to better describe – and understand – economic activity and its interactions with the global environment.

18.1 Introduction

While there is little if any debate among natural scientists and engineers that models of real-world systems must comply with the laws of thermodynamics, the same cannot be said for economists. Questions as to the relevance of thermodynamics, especially the second law, for economic processes have been hotly debated for at least the last three decades and continue to divide the discipline into separate camps (Meshkov and Berry 1979, Young 1991, Daly 1992, Townsend 1992). One of these camps argues that economics is the discipline that addresses how scarce resources can best be used to meet the needs of humans. Chief among these resources are land, labor and human-made capital, though more recently the role of materials, energy, and environmental services such as waste assimilation and absorption, have been acknowledged. Collectively, raw materials, energy and environmental systems are frequently lumped into a category of natural capital (El Serafy 1991, Costanza and Daly 1992) and treated analogously to human-made capital. Application of technology is seen as the means to combine and/or transform land, labor and the various forms of capital into the products that help satisfy human needs. The latter are frequently assumed as given. Markets and other institutions have the goal of revealing to the multitude of decision makers relevant information for the exchange and use of goods and services such that needs are met most (economically) efficiently (Arrow and Hahn 1971, Malinvaud 1972).

An elaborate set of theories and concepts have been developed to show how, for given conditions, market mechanisms can help bring about efficient exchange, allocation and use of land, labor and capital, and whether the resulting prices for inputs into and outputs from the economic process lead to equilibria in consumer and producer markets, maximum satisfaction of consumers, and maximum profits of producers. Where conditions for optimality are not met, interventions are sought to get closer to, or actually achieve economic equilibrium. The main instruments for interventions are changes in relative prices of goods and services, for example by imposing taxes, or changes in the forms of markets themselves, such as by deregulating monopolies or providing incentives for increased competition (Katz and Rose 1998).

Since prices express economic values as determined by the relative scarcities of goods and services, the preferences of consumers, the technologies of producers and the workings of the market place, there is no necessity of economic models to comply with physical laws, such as the laws of thermodynamics. The latter are seen as pertaining to the merely physical world, which is only one of many determinants for the optimal allocation of scarce resources.

The point of departure for the second camp is the recognition that all economic processes do take place in a physical world, that the transformation of inputs into desired and waste products is governed by physical laws, that a set of material, energy and information flows underlies any exchange of value, and that all this takes place within the context of the ecosystem Earth, whose processes are constrained by given endowments of materials as well as stocks and flows of (thermodynamically available) energy. Two interrelated research agendas have developed from this world view (Binswanger 1993). One of these concentrates on process levels, such as material and energy use in production and consumption, and how these may be modeled to be consistent with the laws of thermodynamics. The other research agenda focuses on the macro-scale where issues of larger system constraints, carrying capacity and sustainability play a key role. Both of these areas are largely described by efforts to adopt insights from thermodynamics to guide modeling and decision making, though some are clearly more based on analogies with physical and biological systems, rather than the physical principles themselves.

Instead of trying to settle the debate about the relevance of thermodynamics for economic processes, this chapter lays out some areas on which research has focused so far, the insights that have been generated, and issues that remain unresolved. The following two sections discuss thermodynamic constraints at the process- and macroeconomic levels, respectively. Section 18.4 addresses evolutionary issues and Sect. 18.5 the concepts of information and knowledge in economic modeling. The chapter closes with a brief summary and conclusions.

18.2 Thermodynamic Constraints on Production and Consumption

Thermodynamics sets the limits to the ability of economic systems to generate a desired output. The extents to which wastes accrue in the process and limit economic performance have been most extensively investigated at the level of individual processes (Ruth 1993, 1995a). Early studies have used mass and energy balances to help account for by-products (wastes) which typically are not traced in economic accounts of production and consumption (Ayres and Kneese 1969, Victor 1972, Ayres 1978). Subsequently, using second law analysis, efforts have been made to also reflect changes in the quality of materials and energy as production and consumption take place (Berry and Anderson 1982, Ayres and Nair 1984). These studies typically quantify entropy, negentropy, or exergy changes as a result of economic activity.

Empirical analyses into thermodynamic constraints at the process or technology level have shown where technological improvements exhibit strong diminishing returns due to thermodynamic limits, and where there is substantial room for improvements in the efficiency of energy and material use (Chapman and Roberts 1983). For example, in the area of energy technologies, thermodynamic analyses suggest good reasons to not pursue research on thermal methods for generating hydrogen from water (Warner and Berry 1986). In contrast, thermodynamic analyses provide strong motivation to carry out research on heat-driven physical separation processes (Orlov and Berry 1991).

Comparisons of first law efficiencies with second law efficiencies also highlight the role of the quality of energy in industrial processes and guide the choice among alternatives (Gyftopolos et al. 1974). Detailed exergy analyses identify the locations of energy degradation in a process to improve operation or technology. Among the most comprehensive studies calculating the deviation from the thermodynamic ideal is Szargut et al. (1988). Their study quantified a "cumulative degree of perfection" – a measure of the deviation from thermodynamic ideals of a series of production processes spanning from recovery of raw materials to the refining of the desired products. The closer each step of a production process operates to its thermodynamic ideal and the shorter the chain of process steps, the higher is the cumulative degree of perfection of the entire process. Since production of desirable output from raw materials typically entails long process chains, many of which are far from their thermodynamic ideals, overall low efficiencies result. Similarly, extraction and conversion of energy for generation of electricity to final consumers occurs at single-digit cumulative degrees of perfection. As a consequence, overall second law efficiencies of the US economy is roughly 2% (see, for example, Ayres 1989).

A selection of estimates are presented in Table 18.1. The values range widely, indicating significant opportunities for many material processing industries to increase their material and energy use efficiencies. It is also striking, however, that fossil fuel processing and paper and plastic production show less room for efficiency improvements – products to which modern industrial societies became increasingly accustomed.

Table 18.1. Cumulative Degree of Perfection for the Production of Materials. (Source: Szargut et al. 1988, pp. 180–190)

Material	Cumulative Degree of Perfection [%]
Glass (from raw materials in the ground)	.8
Nickel (from ore concentrate)	1.6
Copper (electrolytic; from Cu_2S ore)	3.1
Lead (from ore in the ground)	3.5
Cement (wet method; medium rotary kiln)	6.2
Tin (from ore concentrate with 20% tin in concentrate)	15.4
Paper (from standing timber)	18.7
Iron Ore Sinter (from ore)	27.0
Cellulose (from wood; waste products used as fuels)	27.4
Open Hearth Steel (liquid; 70% scrap)	34.4
Electric Steel (liquid, 100% scrap)	35.4
Pig Iron (from magnetic taconite at 32.5% fe content)	40.0
Pig Iron (from ore in the ground; mining and blast furnace from high grade haematite ore)	44.0
Polyethylene (low density; from crude oil)	52.5
Benzene (from crude oil)	68.1
Benzene (liquid; from bituminous coal)	71.6
Paper (from waste paper)	74.3
Diesel Oil (typical value)	83.5
Natural Gas (typical value)	87.5

18.3 Constraints at Macroeconomic Levels

Constraints at the macroeconomic level cannot be defined as explicitly as they can for individual processes or industries. The thermodynamic constraints that are imposed on individual production processes may be of little meaning for an industry as a whole because industry is able to choose among various processes. Processes that are close to their thermodynamic ideal can be replaced altogether by those that produce comparable outputs with more direct methods, thus eliminating waste in intermediate production stages (Berg 1980).

Traditional economic models typically assume that such substitution of one process for another, or one kind of input into production for another is always possible. Historically such substitution was key to economic growth and development, and is illustrated, for example, in the case of substitution of machinery and fertilizers for some of the labor and land used in agriculture. Increasingly, however, limits to such substitution are experienced because of larger systems constraints – such as limited availability of resources, and limited waste absorption capacities of ecosystems (Daly 1992).

Extensions of standard (neoclassical) economic growth models to incorporate first and second law constraints on economic and ecosystem processes show, for example, that an economically optimal path leads to a stationary state with finite capital and finite technical knowledge (Ayres and Miller 1980) and indicate upper limits for savings in exergy (Ruth 1993). Similarly, variants of classical economic models fare no better under the light of physical realities. For example, Perrings (1987) develops a variant of the von Neumann-Leontief-Sraffa classical general equilibrium model in the context of a jointly determined economy-environment system subject to a conservation of mass constraint. The model demonstrates that the conservation of mass contradicts the free disposal, free gifts, and non-innovation assumptions of such models. Accounting for the conservation of mass destroys the determinacy of the closed, time variant system in the classical model. An expanding economy causes continuous disequilibrating change in the environment. Since market prices in an interdependent economy-environment system often do not accurately reflect environmental change, such transformations of the environment often will go unanticipated.

On a practical level, increases in local order in the economy are accompanied by decreases in order elsewhere as a result of the degradation of exergy in the fuels used for extractive processes, and the dispersal of waste materials. To reduce, for example, the degradation and dispersal of materials requires increasingly sophisticated technologies to locate raw materials, change their thermodynamic state and trace waste products. Knowledge about the state and fate of waste products is an essential prerequisite for closing material cycles. However, without proper knowledge embodied in technologies and socioeconomic institutions an "industrial ecology" (Jelinski et al. 1992; Graedel and Allenby 1995) based on closed material cycles cannot be achieved.

While first law analyses of production processes provide a quantification of material and energy waste streams from a production process, second law analyses capture the qualitative change that occurs as materials change their thermodynamic state and as the quality of energy decreases. Together, the two help make valuable conceptual and empirical connections between economic and ecosystem processes: characterization of the waste stream in terms of its exergy may be used as a measure of its ability to affect physical, chemical and biological processes in the environment (Ayres et al. 1996). High-exergy (i.e., low entropy per unit energy) waste flows have a high ability to trigger environmental change; waste flows of zero exergy have a material composition, temperature and pressure that is indistinguishable from the environment, and as a result do not affect environmental processes (Ahrendts 1980). However, even though exergy flows have the ability to cause system change, a measure of these flows not necessarily indicates whether and how much a system actually changes when it receives those flows. Some environmental systems, for example, have evolved strategies to cope with specific materials even in high concentrations or in extreme temperature environments while others have not.

To judge the impact of exergy flows on biological structure and function, and thus how release or appropriation of exergy flows affects the larger ecosystems within which economies operate, requires that we know the history and current state of the systems that yield or receive those flows, their ability to cope with changes in exergy flows and the distance these systems are away from thresholds (Ruth and Bullard 1993). The latter is frequently impossible to know a priori. As a consequence, significant efforts in field research and modeling are required to anticipate the impacts of exergy flows and to guide technology choice towards a reduction in resource (exergy) depletion and environmental impact (exergy loading). Yet, even if all the data were available to describe the systems' past and current states, the nonlinearities that underlie feedback mechanisms in the system will make it inherently impossible to predict their future behavior. Even small variation in initial conditions or parameter values may lead to fundamentally different trajectories for the system. The recognition that there are fundamental limits of economic predictability, which has largely come from insights generated in thermodynamics and associated mathematical theories and concepts, has begun to trouble, if not undermine, conventional deterministic economics and has prompted some to adopt a physics-based evolutionary view of economic processes (see, e.g., Tiezzi 2002).

18.4 Thermodynamics and the "Evolution" of Economic Processes

Systems that exchange mass or energy with their surroundings and temporarily maintain themselves in a state away from thermodynamic equilibrium and at a locally reduced level of entropy are called nonequilibrium systems. Biological systems and economic systems fall into this category (Reiss 1994).

Concepts developed in nonequilibrium thermodynamics are particularly suited for the analysis of changes in organization and complexity of open systems. The concept of entropy production, P, is central to nonequilibrium thermodynamics. For an open system, the entropy generated inside the system per unit of time, d_iS/dt, is the sum of the products of the rates J_k and the corresponding n forces $X_k(k{=}1{:}n)$ such that

$$P = \frac{d_iS}{dt} = \sum_{k=1}^{n} J_k X_k > 0 \tag{18.1}$$

(Prigogine 1980). The rates J_k may characterize, for example, heat flow across finite temperature boundaries, diffusion, or inelastic deformation, accompanied by generalized forces X_k such as affinities and gradients of temperatures or chemical potentials. If, and only if, the system is close to equilibrium, then these fluxes can be linearized to their driving fluxes – i.e., $J_k \sim \sum L_{ki} X_i$.

In steady state, the entropy production inside the open system must be accompanied by a net outflow of entropy into the system's surroundings. Systems approaching the steady state are characterized by a decrease in entropy production, i.e., $dP/dt < 0$. Even though originally developed for non-living, physical systems, this "minimum entropy production principle" serves as a cornerstone in many thermodynamic analyses of living systems (Johnson 1981, Gladyshev 1982, Schneider 1988).

Far from equilibrium, it is typically not appropriate to assume the linearity suggested above. There is considerable debate, in how far the "close-to-equilibrium" assumption can be maintained for the analysis of production and consumption processes in living systems, engineering systems or economies. In the nonlinear realm, systems can be characterized by a generalized potential, the excess entropy production. Excess entropy production is defined as

$$P_E = \sum_k^n \delta J_k \, \delta X_k \tag{18.2}$$

with δJ_K and δX_K as deviations from the values J_k and X_k at the steady state. Unlike for systems in, or close to, equilibrium, the sign of P_E is generally not well defined. However, close to equilibrium the sign of P_E is equal to that of P (Glansdorff and Prigogine 1971).

Calculations of excess entropy production show decreases in P_E in open systems moving towards steady state. For example, an increase in temperature gradients in a far-from-equilibrium system may trigger increasingly complex structures, as Bénard's experiments clearly demonstrate (Prigogine 1980). The evolution towards these structures is typically not smooth but accompanied by discontinuities and instabilities. In the critical transition point between stability and instability, a more complex structure emerges and $P_E = 0$ (Glansdorff and Prigogine 1971). Excess entropy production can therefore serve as a measure of changes in the structure and stability of a system. However, its applicability to real, living systems and especially to economic systems is severely limited by the lack of data sufficient for meaningful calculations (Ruth 1995b), and the bulk of non-equilibrium thermodynamic "analyses" of economic systems to date has been guided by analogies or metaphors, and exhibit little theoretical and empirical rigor.

18.5 Information and Knowledge

When physical principles are applied to living systems, the systems typically are analyzed at either very fine or very coarse resolutions. Note that concrete judgments about the demise of individuals or large assemblages cannot be made by using thermodynamics alone. Yet, intermediate levels of organization, such as households and firms, are the most tangible in economic decision-making. Furthermore, the evolutionary concepts derived from physics are inapt in accommodating the conscious influences of humans that take place at that level. Thus, it seems prudent at the current stage of development of the concepts to apply thermodynamics to larger entities in an economic system or entire economies than, for example, to decision making of individuals (Ruth 1996).

Once the appropriate level of analysis has been identified, the question becomes how to measure and assess evolutionary change (O'Connor 1991). For physical and biological systems, measurement is typically done in terms of exergy flows across system boundaries. For example, measurement of exergy fluxes into and out of old growth forests in comparison to less complex ecosystems, such as plantations or barren ground, indicate that indeed the more evolved ecosystems possess a higher ability to degrade the gradients imposed on them (Luvall and Holbo 1991). Similarly, exergy-flux based studies can be done for socioeconomic systems, as has been illustrated by Hannon et al. (1993) for the case of industrial and Amish agriculture. Judging the possible direction of evolutionary change in an economy, from these exergy-based studies however, is difficult.

It is not only the change in the quantity and quality of inputs into economic processes that is fundamental to the evolution of the economy, but the fact that economic change feeds on itself. Can thermodynamics provide measures of change in material and energy inputs and economic output that are indicative of evolutionary processes? At what resolution is such a quantification valid and meaningful?

The concepts of information and knowledge can be defined from a thermodynamic perspective and linked to the material and energy sources used in an economy (Ayres 1994). They are general enough to encompass changes in output quality, technology, even changes in socio-economic characteristics that are concomitant with economic evolution.

The relation between energy and information is fundamental in thermodynamics, having served to refute Maxwell's demon (Maxwell 1867, Leff 1990), and thus to confirm the validity of the second law of thermodynamics (Szilard 1929). The information concept has been adopted by Shannon (1948), Shannon and Weaver (1949) and Wiener (1948) in the context of communication theory. Information was defined in communication theory as a measure of uncertainty that caused an adjustment in probabilities assigned to a set of answers to a given question (Young 1971). Evans (1969) and Tribus and McIrvine (1971) formalized the connection between Shannon's work and thermodynamic information. Systems with lower entropy are more distinguishable from a reference environment by an observer, and thus, are able to convey more information to that observer than systems in equilibrium with their environment.

These insights have begun to inform the analysis of production and consumption processes (Ayres and Miller 1980; Berg 1980; Chen 1992; Chen 1994; Spreng 1993; Ruth 1995a) and of economy-environment interactions (Ayres et al. 1996; Ruth 1995b). Common to all these studies is the recognition of the fundamental role the concept of information can play in explaining the state and evolution of biological and economic activity.

Making the connection between these thermodynamics and information-based approaches to changes in knowledge, appropriately defined, remains a challenge (Eriksson et al. 1987, Ayres 1994). Some have argued that as knowledge increases, so does the ability to utilize a variety of materials and energy sources, and to produce increasingly specialized products from them. As a result, the ability to actively cause change in the physical, but also biotic and socioeconomic environment of the simple society, is enhanced. Trying to understand the rise and fall of societies in relationship to material and energy use, complexity of production and distribution of goods and services – in short, a society's knowledge – has a long history in anthropology (Cronon 1983, Debeir et al. 1986, Tainter 1988, Perlin 1989). Relating socioeconomic changes to the exchange of materials, energy and information in and among economies, and their relationship to knowledge, may seem a worthwhile endeavor for economics, which will clearly benefit from a more rigorous infusion of concepts and tools from thermodynamics and information theory (c.f. Allen and Lesser 1991).

18.6 Conclusion

This chapter has laid out various avenues through which concepts from thermodynamics have entered economic analysis – from individual production and consumption processes to larger macroeconomic levels. Some of these analyses build on the first and second laws to identify physical limits to economic processes and assess the contribution of material, energy and information flows to changes in the economy and the ecosystem within which it operates. Others explore more broadly the implications that insights from non-equilibrium thermodynamics and complex systems theory have for the behavior and descriptions of economic systems. Both are closely related and juxtaposed to traditional economic modeling where little regard is given to the fundamental physical principles that underlie all economic processes, and the constraints that are imposed on economic growth and development by the interactions between the economy and its physical and biotic environment. Yet, if economics is indeed concerned with efficient allocation of scarce means to meet human needs, then finding a "global optimum" (both in the mathematical and life-support sense) will require that descriptions of individual processes adequately reflect physical reality and that broader systems constraints on overall economic performance are accounted for.

References

Ahrendts J (1980) Reference States, Energy 5: 667–677.
Allen PM, Lesser M (1991) Evolutionary Human Systems: Learning, Ignorance and Subjectivity, in PP Saviotti and JS Metcalfe (eds.) Evolutionary Theories of Economic and Technological Change, Harwood Academic Publishers, Chur, Switzerland, 160–171.
Arrow KJ, F. Hahn F (1971) General Competitive Analysis, Holden-Day, San Francisco.
Ayres RU (1978) Resources, Environment, and Economics: Applications of the Materials/Energy Balance Principle. John Wiley and Sons, New York,
Ayres RU (1989) Energy Efficiency in the US Economy: A New Case for Conservation, International Institute for Applied Systems Analysis, Laxenburg, Austria.
Ayres RU (1994) Information, Entropy, and Progress: A New Evolutionary Paradigm, American Institute of Physics, New York.
Ayres RU, Kneese AV (1969) Production, Consumption, and Externalities, American Economic Review 59: 282–298.
Ayres RU, Miller SM (1980) The role of technical change. Journal of Environmental Economics and Management 7: 353–371.
Ayres RU, Ayres LW, Martinás K (1996) Eco-Thermodynamics: Exergy and Life Cycle Analysis, INSEAD Working Papers, No. 96/04/EPS, INSEAD, Fontainebleau.
Ayres R, Nair I (1984) Thermodynamics and economics 35: 62–71.
Berg CA (1980) Process Integration and the Second Law of Thermodynamics: Future Possibilities, Energy 5: 733–742.
Berry RS, Anderson B (1982) Thermodynamic Constraints in Economic Analysis. In: Schieve WC, Allen PM. (eds.) Self Organization and Dissipative Structures: Applications in the Physical and Social Sciences. University of Texas Press.
Biswanger M (1993) From microscopic to macroscopic theories: entropic aspects of of ecological and economic processes. Ecological Economics 8: 209–234.
Chapman PF, Roberts F (1983) Metal Resources and Energy. Butterworths, London.
Chen X (1992) Substitution of Information for Energy in the System of Production, ENER Bulletin, July 1992, 45–59.
Chen X (1994) Substitution of Information for Energy, Energy Policy 22: 15–27.
Costanza R, Daly HE (1992) Natural capital and sustainable development. Conservation Biology 6: 37–46.
Cronon W (1983) Changes in the Land: Indians, Colonists, and the Ecology of New England, Hill and Wang, New York.
Daly HE (1992) Is the Entropy Law Relevant to the Economics of Natural Resource Scarcity? – Yes, of Course It Is! Journal of Environmental Economics and Management 23: 91–95.
Debeir J-C, Deléage J, Hémery D (1986) In the Servitude of Power: Energy and Civilization Through the Ages, Zed Books, Ltd., London.
El Serafy S (1991) The Environment as Capital, in R. Costanza (ed.) Ecological Economics: The Science and Management of Sustainability, Columbia University Press, New York, 168–175.
Eriksson K-E, Lindgren K, Månsson BÅ (1987). Structure, Context, Complexity, Organization, World Scientific Publishing Co., Singapore.

Evans RB (1969) A Proof that Essergy is the only Consistent Measure of Potential Work, Ph.D.. Thesis, Dartmouth College, Hannover, New Hampshire.
Gladyshev GP (1982) Classical Thermodynamics, Tandemism and Biological Evolution, Journal of Theoretical Biology 94: 225–239.
Glansdorff P, Prigogine I (1971) Thermodynamic Theory of Structure, Stability and Fluctuations, Wiley-Interscience, New York.
Graedel TE, Allenby BR (1995) Industrial Ecology, Prentice Hall, Englewood Cliffs, New Jersey.
Gyftopoulos EP, Lazaridis LJ, Widmer TF (1974) Potential fuel effectiveness in industry. Ballinger Publishing Company, Cambridge, Massachusetts
Hannon B, Ruth M, Delucia E (1993) A Physical View of Sustainability, Ecological Economics 8: 253–268.
Jelinski LW, Graedel TE, Laudise RA, McCall DW, Patel CKN (1992) Industrial Ecology: Concets and Approaches. Proceedings of the National Academy of Science 89: 793–797.
Johnson L (1981) The Thermodynamic Origin of Ecosystems. Canadian Journal of Aquatic Sciences 38: 571–590.
Katz ML, Rose HS (1998) Microeconomics, 3rd Edition, Irwin McGraw-Hill, Boston.
Leff HS (1990) Maxwell's Deamon, Power, and Time, American Journal of Physics 58: 135–141.
Luvall JC, Holbo HR (1991) Thermal Remote Sensing Methods in Landscape Ecology, in M Turner and RH Gardner (eds.) Quantitative Methods in Landscape Ecology, Springer-Verlag, New York, 127–152.
Malinvaud E (1972) Lectures on Microeconomic Theory, North-Holland Publishing Company, Amsterdam, London.
Maxwell JC (1967) Letter to Peter Guthrie Tait, in C.G. Knott, Life and Scientific Work of Peter Guthrie Tait, Cambridge University Press, London, 213–215.
Meshkov N, Berry RS (1979) Can Thermodynamics Say Anything about the Economics of Production? In: Fazzolare RA, Smith CB (eds), Changing Energy Use Futures. Pergamon Press, New York, 374–382.
O'Connor M (1991) Entropy, structure, and organisational change. Ecological Economics 3: 95–122.
Orlov VN, Berry RS (1991) Estimation of minimal heat consumption for heat-driven separation processes via methods of finite-time thermodynamics. Journal of Physical Chemistry 95: 5624–5628.
Perlin J (1989) A Forest Journey: The Role of Wood in the Development of Civilization, Norton, New York.
Perrings C (1987) Economy and Environment: A Theoretical Essay on the Interdependence of Economic and Environmental Systems. Cambridge University Press, Cambridge, New York.
Prigogine I (1980) From Being to Becoming: Time and Complexity in the Physical Sciences, W.H. Freeman and Company, New York.
Reiss JA (1994) Comparative Thermodynamics in Chemistry and Economics, in Burley P, Foster J (eds) Economics and Thermodynamics: New Perspectives on Economic Analysis, Kluwer Academic Publishers, Dortrecht, The Netherlands, 47–72.
Ruth M (1993) Integrating Economics, Ecology, and Thermodynamics. Kluwer Academic, Dordecht, 251 pp.

Ruth M (1995a) Information, order and knowledge in economic and ecological systems: implications for material and energy use. Ecological Economics 13: 99–114

Ruth M (1995b) Thermodynamic implications for natural resource extraction and technical change in U.S. copper mining. Environmental and Resource Economics 6: 187–206

Ruth M (1996) Evolutionary Economics at the Crossroads of Biology and Physics, Journal of Social and Evolutionary Systems 19: 125–144.

Ruth M, Bullard CW (1993) Information, Production, and Utility. Energy Policy 21(10): 1059–1066.

Schneider ED (1988) Thermodynamics, Ecological Succession and Natural Selection, in Weber B, Depew D (eds), Entropy, Information and Evolution, MIT Press, Cambridge, Massachusetts.

Shannon CE (1948) A Mathematical Theory of Communications, *Bell Systems Technology Journal*, Vol. 27.

Shannon CE, Weaver W (1949) The Mathematical Theory of Information, University of Illinois Press, Urbana.

Spreng DT (1993) Possibilities for Substitution Between Energy, Time and Information, Energy Policy, January 1993, 13–23.

Szargut J, Morris DR, Steward FR (1988) Exergy Analysis of Thermal, Chemical, and Metallurgical Processes, Hemisphere Publishing Corporation, New York.

Szilard L (1929) Über die Entropieverminderung in einem Thermodynamischen System bei Eingriffe Intelligenter Wesen. Zeitschrift für Physik 53: 840–960.

Tainter JA (1988) The Collapse of Complex Societies, Cambridge University Press, Cambridge.

Tiezzi E (2002) The Essence of Time, WIT Press, Billerica, Massachusetts.

Townsend KN (1992) Is the Entropy Law Relevant to the Economics of Natural Resource Scarcity? Comment, Journal of Environmental Economics and Management 23: 96–100.

Tribus M, McIrvine EC (1971) Energy and Information, Scientific American 225: 179–188.

Victor PA (1972) Pollution: Economy and Environment, Allen and Unwin, London.

Warner JW, Berry RS (1986) Hydrogen separation and the direct high-temperature splitting of water. International Journal for Hydrogen Energy 11: 91–100.

Wiener N (1948) Cybernetics, John Wiley, New York.

Young JF (1971) Information Theory, Wiley Interscience, New York.

Young JT (1991) Is the Entropy Law Relevant to the Economics of Natural Resource Scarcity? Journal of Environmental Economics and Management 21: 169–179.

Index